Aero Engine Combustor Casing

Experimental Design and Fatigue Studies

Aero Engine Combustor Casing

Experimental Design and Fatigue Studies

Sashi Kanta Panigrahi and Niranjan Sarangi

CRC Press
Taylor & Francis Group
Boca Raton London New York

CRC Press is an imprint of the
Taylor & Francis Group, an **informa** business

CRC Press
Taylor & Francis Group
6000 Broken Sound Parkway NW, Suite 300
Boca Raton, FL 33487-2742

First issued in paperback 2020

ISBN 13: 978-0-367-57353-9 (pbk)
ISBN 13: 978-1-138-03283-5 (hbk)

Library of Congress Cataloging-in-Publication Data

Names: Panigrahi, Shashi Kanta, author | Sarangi, Niranjan, 1946– author.
Title: Aero Engine Combustor Casing : Experimental Design and Fatigue Studies/ Shashi Kanta Panigrahi, Niranjan Sarangi.
Description: Boca Raton : Taylor & Francis, a CRC title, part of the Taylor & Francis imprint, a member of the Taylor & Francis Group, the academic division of T&F Informa, plc, [2017] | Includes bibliographical references and index.
Identifiers: LCCN 2016048815| ISBN 9781138032835 (hardback : alk. paper) | ISBN 9781138032842 (ebook)
Subjects: LCSH: Aircraft gas-turbines–Combustion chambers–Design and construction. | Aircraft gas-turbines–Combustion chambers–Materials–Fatigue.
Classification: LCC TL709.5.C55 P36 2017 | DDC 629.134/353–dc23
LC record available at https://lccn.loc.gov/2016048815

Visit the Taylor & Francis Web site at
http://www.taylorandfrancis.com

and the CRC Press Web site at
http://www.crcpress.com

For Product Safety Concerns and Information please contact our EU representative:
GPSR@taylorandfrancis.com
Taylor & Francis Verlag GmbH, Kaufingerstraße 24, 80331 München, Germany.

To our parents

and

loving family members

Contents

Preface xi
Authors xiii

1. Introduction 1
1.1 Background and Motivation 1
1.2 Outline of this Book 2
1.3 Initial Phase of Aero Engine Development 4
1.4 Working Principle of Aero Gas Turbine Engines 6
1.5 Critical Components of an Aero Engine 7
1.6 Aero Engine Compressor 7
1.7 Aero Engine Combustion Chamber 10
1.8 Aero Engine Turbine 10
1.9 Aero Engine Propelling Nozzle 12
1.10 Design Philosophy of an Aero Engine Combustion Chamber 14
1.11 Types of Combustion Chamber in Aero Engines 15
1.11.1 Multiple Combustion Chambers 15
1.11.2 Tubo-Annular Combustion Chamber 16
1.11.3 Annular Combustion Chamber 16
1.12 Complexities in Combustion Chamber Design 17
1.13 Materials Used for Combustion Chamber 19
1.14 Significant Contributions 20
1.15 Summary 21

2. Fatigue Design Philosophy of an Aero Engine Combustor Casing 23
2.1 Introduction 23
2.2 Combustion Chamber Design 23
2.3 Fatigue Failure in Aero Engines 25
2.4 Fatigue Cycle Counting Methods 29
2.5 Fatigue Life Evaluation Methods 34
2.6 Fatigue Damage Assessment 37
2.7 Nondestructive Testing Methods 39
2.7.1 Radiography Inspection 40
2.7.2 Magnetic Particle Inspection 42
2.7.3 Dye Penetrant Inspection 44
2.7.4 Ultrasonic Inspection 45
2.7.5 Eddy Current and Electromagnetic Inspection 47
2.8 Summary of the Design Philosophy 49
2.9 Important Design Considerations for Combustor Casing 50
2.10 Summary 51

3. Development of Test Facility and Test Setup 53
3.1 Introduction 53
3.2 Airworthiness and Certification 55
3.3 Description of the Test Facility and Its Subsystems 56
3.3.1 Hydraulic Power Supply 57
3.3.2 Servo Valve with Manifold 58
3.3.3 Safety Relief Valve 59
3.3.4 Servo Controller 59
3.3.5 Pressure Transducer 59
3.4 Integration of the Subsystems 60
3.5 Design and Manufacturing of Adaptors 61
3.5.1 Bottom Fixing Plate 62
3.5.2 Bottom Sealing Drum 62
3.5.3 Inner Bottom Fixing Plate 63
3.5.4 Special Tie Rod 64
3.5.5 Top Sealing Plate 65
3.5.6 Top Fixing Plate 65
3.5.7 Top Support Plate 65
3.5.8 Top Holding Plate 66
3.5.9 Special Studs 67
3.5.10 Seals 68
3.6 Summary 68

4. Manufacturing of an Aero Engine Combustor Casing, the Experimental Evaluation of Its Fatigue Life, and Correlation with Numerical Results 69
4.1 Introduction 69
4.2 Manufacturing Method of the Combustor Casing 70
4.2.1 Metal Spinning Process 70
4.2.1.1 Advantages of the Metal Spinning Process 71
4.2.1.2 Disadvantages of Metal Spinning Process 72
4.2.2 Electron Beam Welding Method 72
4.2.3 Tungsten Inert Gas Welding Method 74
4.3 Configuration of the Combustor Casing 75
4.4 Experimental Evaluation of Fatigue Life 76
4.4.1 Instrumentation 77
4.4.2 Inspection Methodology 79
4.5 Mechanical Properties of Combustor Casing Material 81
4.6 Numerical Analysis of the Combustor Casing 82
4.6.1 Details of the Finite Element Model 83
4.6.2 Finite Element Model Quality Parameters 84
4.6.3 Boundary Conditions 85
4.6.4 Load 85
4.6.5 Pressure Load 85
4.6.6 Thermal Load 85

4.6.7 Nusselt Equation ... 87
4.6.8 Dittus Boelter Equation ... 87
4.7 Numerical Analysis of the Igniter Boss and Its Correlation with Experimental Data ... 89
4.8 Results and Discussions ... 98
4.9 Summary ... 98

5. Reassessment of Fatigue Life of the Modified Combustor Casing ... 101
5.1 Introduction ... 101
5.2 Modified Manufacturing Methodology ... 101
5.2.1 Forging Process: Benefits and Drawbacks ... 101
5.3 Instrumentation of the Modified Combustor Casing ... 106
5.4 Assembly and Trial Run ... 106
5.5 Testing ... 109
5.6 Results and Discussion ... 109
5.7 Summary ... 110

6. Safety Test on Modified Combustor Casing ... 113
6.1 Introduction ... 113
6.2 Test Component Details ... 114
6.3 Instrumentation and Testing ... 115
6.4 Results and Discussion ... 117
6.5 Summary ... 118

7. Effect of Fatigue on the Proof Strength of an Aero Engine Combustor Casing ... 119
7.1 Introduction ... 119
7.2 Metallographic Techniques in Failure Analysis ... 120
7.2.1 Stereo Zoom Microscopy ... 121
7.2.2 Scanning Electron Microscopy ... 121
7.2.3 Optical Microscopy ... 125
7.3 Details of the Test Component ... 126
7.4 Experimental Procedure ... 127
7.5 Observation ... 128
7.5.1 Visual Observation ... 128
7.5.2 Stereo Zoom Microscopy ... 128
7.5.3 Scanning Electron Microscopy ... 129
7.5.3.1 Fractography ... 129
7.5.3.2 Energy Dispersive X-Ray Analysis ... 131
7.5.4 Optical Microscopy ... 131
7.5.5 Microhardness Measurement ... 131
7.6 Discussions ... 137
7.7 Summary ... 137

8. Conclusions 139
8.1 Certification and Acceptance of the Combustor Casing 139
8.2 Specific Conclusions 139
8.3 Scope for Further Work 140

References 143

Index 153

Preface

The combustion chamber in an aero engine is where the incoming air stream from the compressor is heated by the addition and combustion of fuel. A combustion chamber must therefore satisfy many conflicting requirements:

- It must be capable of initiating the easy ignition of fuel and must operate stably under a wide range of conditions.
- The fuel in the chamber must undergo complete combustion while producing the minimum of pollutants.
- To avoid damage to the turbine of the aero engine, the mixture of air and fuel must achieve an acceptable temperature distribution of exhaust gases.
- The combustor must operate with minimum pressure loss to maintain high overall performance.

To achieve these functionalities together with an acceptable operating life while keeping size, weight, and cost to a minimum represents a considerable design challenge.

It is widely accepted that there is limited literature available in the public domain on the methodology for the life evaluation of a combustion chamber due to the technologies of the various aero engine developers being classified. The only reliable approach through which to determine the fatigue life of a combustion chamber is to design and develop experiments that replicate operational conditions and that are capable of verifying whether an assessment of a combustion chamber life evaluation is correct.

The development of such a test facility is a complex matter as it must respond sensitively and provide detailed results. Various companies have used different approaches. Some industries tried to replicate engine conditions using hot gases; this approach was not only costly, but also very difficult to operate. Other test facilities have been based on pressure cycling; these are technically more convenient and yield better results.

Over recent years, different types of combustor configuration have been widely used in existing aero engines. Of them, the annular combustor is prevalent in modern aero engine designs and is used prolifically in a number of industrial applications. The requirements of a combustor chamber for an aero engine are the provision of high combustion efficiency, sufficient fatigue strength, maximum durability, better reliability and smooth ignition, good stable operation at varied pressures and air/fuel ratios, and a long operational life.

To address all the issues and challenges pertaining to the design, analysis, and development of a combustor chamber for an aero engine, this book discusses a variety of aspects. Among others, these include

- Theoretical investigation
- Thermodynamic processes in a combustion chamber
- Mechanical and manufacturing design consideration and stress analysis
- Various life evaluation methods
- The performance of a combustion chamber
- An experimental approach to determining the fatigue life of a chamber
- Fatigue life testing of a combustion chamber
- Analysis of test data

The focus of this book is on all aspects of experimentation, the development of a test facility, manufacturing procedures, and the analytical and experimental study of fatigue life prediction for an aero engine combustor casing. Many mathematical expressions have been put in a clear and straightforward way and, usually, do not require knowledge of a more mathematical nature. A large number of research articles are provided at the end of this book and provide the opportunity for further reading. The authors have also provided a considerable number of images associated with combustion chambers and hope that those involved with the topic—students, academicians, researchers, industry professionals, and scientists—will take advantage of these images in their work in this field. The authors take this opportunity to thank all their friends and family members who assisted in the preparation of this book. They are also indebted to CRC Press for reviewing and accepting this book for publication.

Sashi Kanta Panigrahi
Niranjan Sarangi

Authors

Dr. Sashi Kanta Panigrahi (PhD, IIT Kharagpur) is working as dean (SA) and professor in the Department of Mechanical Engineering of Defence Institute of Advanced Technology (under DIAT), Pune, India. Before he joined DIAT, he was working in many other universities/Institutions within the country. He has also worked as an international visiting academic with the University of New South Wales at the Australian Defence Force Academy (UNSW@ADFA). He has a wide and intensive teaching, research, training, and administrative experience after completing his PhD from Indian Institute of Technology, Kharagpur. He obtained his bachelor's degree and master's degree in mechanical engineering from Veer Surendra Sai Technical University (VSSUT), formerly known as University College of Engineering (UCE), Burla, and National Institute of Technology (NIT), Rourkela. His research works are primarily in the areas of analysis and design of composite materials, characterization of FRP composite materials, finite-element analysis of FRP composite materials and composite structures, natural fiber reinforced composite (NFRC) materials, fracture mechanics principle applicable to modeling and simulation of damages in orthotropic and isotropic materials, and material characterization/stress analysis/solid mechanics/machine design. He has been working on the development of advanced finite-element methods and nonlinear finite-element modeling and analyses of engineering structures with functionally graded/monolithic adhesively bonded joints. He has published over 150 research articles in peer-reviewed scholarly research papers international journals/conferences, including 2 books/monograph/conference proceedings/many lecture materials. He has been awarded with "Bharat Jyoti" in the year 2012 and conferred with fellow, Institution of Engineers, India, and fellow, Indian Society of Mechanical Engineering. Recently, he has been awarded with "Distinguished Scientist" by Venus International Research Foundation, India. Many PhD and MTech students have already been graduated under his direct guidance/supervision. He is a member of many international/national professional bodies and has been a frequent reviewer for many leading peer-reviewed international journals of high standards. He has also served as technical committee member or advisory board member in several national/international conferences.

Niranjan Sarangi received his first class BE in mechanical engineering in 1992 from University College of Engineering, Burla, Odisha, India. He completed his MTech with specialization in production engineering in the Mechanical Engineering Department at the Institute of Technology, Banaras Hindu University (IT-BHU), Varanasi, Uttar Pradesh, India in July 1995. From August 1995 to 1996, he was a faculty member of the Mechanical Engineering Department of Gangamai College of Engineering, Dhule, Maharashtra, India. In September 1996, he joined the Institute of Armament Technology (IAT) of the Defence Research and Development Organisation (DRDO), Pune, Maharashtra, India, as Scientist-B. He was posted to the Gas Turbine Research Establishment (GTRE) of DRDO in Bangalore in March 1997 and continues there as a scientist. At present, he is working as a Scientist-F in GTRE. During his tenure at GTRE, he has been involved in the structural integrity testing of critical aero engine components for their design validation and airworthiness certification. His research interests also include the evaluation of the fatigue life and safety capability of aero engine components for their airworthiness certification.

Between 2007 and 2013, he was engaged in PhD dissertation "Fatigue life evaluation of an aero engine combustor casing" at the Defence Institute of Advanced Technology (deemed university), Pune, Maharashtra, India, while in service at GTRE, Bangalore. He has published good-quality research papers in peer-reviewed international journals and has presented several research papers at various national and international conferences.

1

Introduction

1.1 Background and Motivation

The main objective of an aero gas turbine engine is to provide thrust. To perform this duty, the engine must meet weight and efficiency targets. Of the thousands of parts comprising an aircraft gas turbine engine, the most critical parts that have the most significant effect on thrust are the rotating components and the high-pressure casings, such as the combustor casing. The first step in the process of establishing and minimizing the risk of failure of aircraft engines is to perform a failure mode and effect analysis to identify the possible consequences of failure of an individual part and of multiple failures in service. From this analysis, the components are categorized as "critical" or "noncritical" depending on the consequences of their failure.

The combustor casing is classified as one of the most critical components as it experiences the highest pressure in the aero engine and its failure can have catastrophic effects, resulting in the destruction of the engine due to uncontained fire from the flame tube. Therefore, the combustor casing should be designed with utmost care and include various design features such as circumferential welds, adaptors, and bosses for the ignitor, customer bleed, pressure probes, atomizers, and boroscope ports, and so on. Mounting of these bosses by welding over the double-curvature casing surface acts as stress raisers and reduces the strength and fatigue life of the component.

For any critical aero engine component, it is a mandatory requirement that the component must demonstrate its strength and fatigue life in a simulated engine environment before being declared fit for use in the engine. To meet this requirement, a full-scale prototype of the component needs to be tested in a test facility designed to simulate the engine environment. Development of the test facility to test the component in the engine environment is a challenging task. The necessary adaptors are designed to adapt the component suitably in the test facility and to apply the load that the component experiences in the engine. Being a highly critical component, a combustor's design validation test is undertaken to meet the airworthiness requirement of the engine. The primary motivation of this book is derived from this requirement.

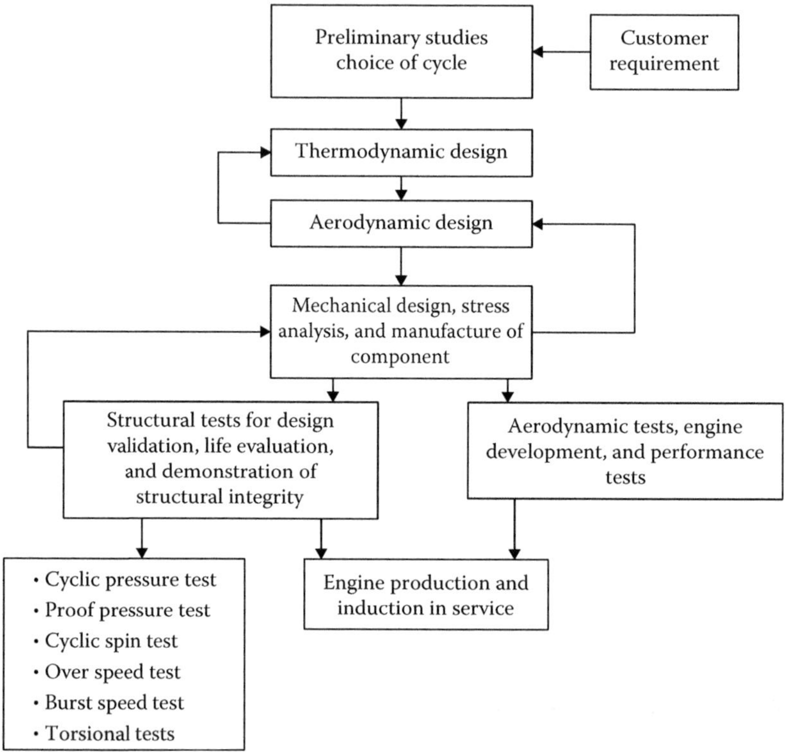

FIGURE 1.1
Importance of structural tests in an aero engine development cycle.

This book aims at providing an expansive framework for evaluation of the fatigue life of a full-scale model of an aero engine combustor casing, identifying critical parameters affecting its life, and recommending suitable design modifications for life improvement. The importance of the structural test of a critical component such as a combustor casing is outlined in Figure 1.1.

1.2 Outline of this Book

Figure 1.2 outlines the important sections of the research work that constitute this book. The details of the research, implementation of the outcomes, and their demonstrations are presented in eight chapters. In order to provide an easier and streamlined reading of this book, each chapter is written as an independent paper that presents an introduction, methodology, test details (if any), and conclusions.

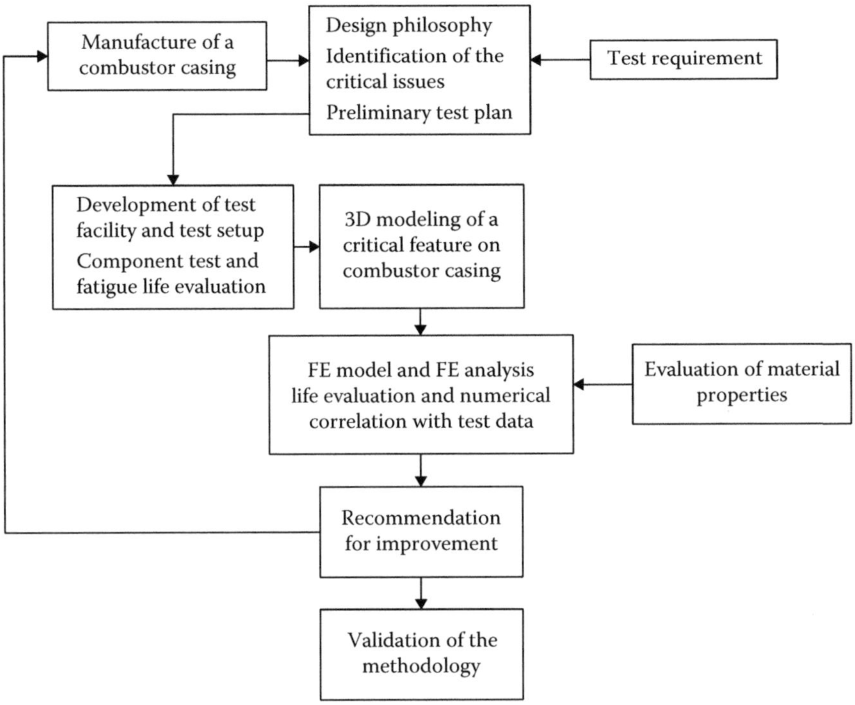

FIGURE 1.2
Outline of this book.

After this introductory chapter, which presents the development sequences of aero engines and aero engine combustion chambers, followed by the significant contributions made in this study, Chapter 2 provides the fatigue design philosophy of an aero engine combustor casing, covering the research work carried out in the fields related to the present publication, the theoretical and experimental details of fatigue life evaluation of a combustor casing, and offering an overview of life evaluation methodologies and relevant technical literature.

Chapter 3 provides the details pertaining to the development of a novel servo hydraulic pressure test rig to carry out the fatigue test on the combustor casing. Design and manufacture of the adaptors and the test setup required to assemble the combustor casing in the test facility are also highlighted in this chapter.

Chapter 4 presents the manufacturing details of an aero engine combustor casing and its experimental evaluation by conducting a cyclic pressure test is explained. When the component failed prematurely, it emphasized the need to focus on the critical design feature that led to its failure. This chapter emphasizes the methodology followed to identify the material properties, the numerical analysis of the combustor casing, and the critical feature using the finite element method and the life evaluation methodology. This information was correlated with the values obtained during testing in order to recommend design and manufacturing modifications of the component.

In accordance with the recommendations in Chapter 4, a modified combustor casing was manufactured by machining forging forged billet. The details of this manufacturing process are explained in Chapter 5. In addition to this, the instrumentation details carried out on this new modified component are explained. This chapter explains the assembly procedure and details of the fatigue life test carried out on the modified combustor casing. The component successfully demonstrated its intended life.

Chapter 6 presents a novel approach to conducting the proof pressure test on the modified combustor casing to ensure its overpressure capability during its operation. The component withstood the proof pressure, which is double the maximum operating pressure applied during the cyclic pressure test.

To enhance the study further, it was planned to evaluate the effect of fatigue on the proof strength of the combustor casing by applying proof pressure on a fatigue loaded combustor casing. The study indicates that proof strength is affected by fatigue. Details of this study are presented in Chapter 7.

Chapter 8 presents a comprehensive summary emphasizing the major achievements, experimental results, and the conclusions drawn.

1.3 Initial Phase of Aero Engine Development

During the period between the two World Wars, the design of aircraft engines improved dramatically and made possible unprecedented progress in aircraft design. Engine development in those days—and, to a large extent, even today—is a very laborious process. The detailed process of building an engine, running it to destruction, analyzing what failed, and redesigning the necessary parts is a repetitive process. No product ever comes to market without a number of engineers having spent many hours perfecting that product. This is especially true for aircraft engines which, by their very nature, push all the limits of ingenuity, materials, and manufacturing processes.

Since the late 1930s, enormous achievements have been witnessed in the field of aero engines, covering technology, design, and manufacture, advancing the state of the art not only for gas turbines, but also for many related industries and products. Gas turbine technology continues to be at the forefront of mechanical and aero technologies, materials and coatings, and production and manufacturing science. The efforts of some of the world's most successful corporations and most respected engineers have placed this industry where it is rightly seen as one of mankind's major engineering achievements.

Recently, Andrew [1] called to mind Sir Frank Whittle who, as a young man in the 1920s, had a vision of aircraft flying above the weather. This vision led Sir Frank Whittle to design the jet engine, a variant of the gas turbine; in January 1930, Sir Frank Whittle [2,3] took out British patent no. 347206. Whittle's first jet engine ran on April 12, 1937. This Whittle engine formed the basis of the modern gas turbine engine.

In 1935, Hans von Ohain started work on a similar design in Germany, apparently unaware of Whittle's work. His first device was strictly experimental and could only run under external power, but he was able to demonstrate the basic concept. Erich Warsitz, a German test pilot, [4] was then introduced to Ernst Heinkel, one of the larger aircraft industrialists of the day, who had immediately seen the promise of Ohain's design. Heinkel had recently purchased the Hirth engine company, and Ohain and his master machinist, Max Hahn, were set up there as a new division of the Hirth company. They had their first HeS 1 centrifugal engine running by September 1937. Unlike Whittle's design, Ohain used hydrogen as fuel, supplied under external pressure. Their subsequent designs culminated in the gasoline-fueled HeS 3 of 1100 lbf (5 kN), which was fitted to Heinkel's simple and compact He 178 airframe and flown by Erich Warsitz in the early morning of August 27, 1939, from Rostock-Marienehe aerodrome; this had been an impressively short time for development. The He 178 was the world's first jet plane.

Austrian Anselm Franz of Junkers' engine division (*Junkers Motoren* or "Jumo") introduced the axial-flow compressor in their jet engine. Jumo was assigned the next engine number in the RLM 109-0xx numbering sequence for gas turbine aircraft powerplants—004—and the result was the Jumo 004 engine. After many lesser technical difficulties were solved, mass production of this engine started in 1944 as a power plant for the world's first jet fighter aircraft, the Messerschmitt Me 262 (and, later, the world's first jet-bomber aircraft, the Arado Ar 234). A variety of reasons conspired to delay the engine's availability, causing the fighter to arrive too late to improve Germany's position in World War II. Nonetheless, it will be remembered as the first use of jet engines in service. Meanwhile, in Britain the Gloster E28/39 had its maiden flight on May 15, 1941 and the Gloster Meteor finally entered service with the Royal Air Force in July 1944.

Following the end of World War II, the German jet aircraft and jet engines were extensively studied by the victorious allies and contributed to work on early Soviet and U.S. jet fighters. The legacy of the axial flow engine is seen in the fact that practically all jet engines on fixed wing aircraft have to some extent been inspired by this design.

By the 1950s, the jet engine was almost universal in combat aircraft; the exceptions were cargo, liaison, and other specialty types of aircraft. By this point, some of the British designs had already been cleared for civilian use and had appeared in early models such as the de Havilland Comet and Avro Canada Jetliner.

By the 1960s, all large civilian aircraft were also jet powered, leaving the piston engine in low-cost niche roles such as cargo flights.

Turbojet engines were still less fuel-efficient than piston engines but, by the 1970s, the advent of high bypass turbofan jet engines, an innovation not foreseen by early commentators such as Buckingham [5], fuel efficiency at high speeds and high altitudes was about the same as the best piston and propeller engines.

At present, gas turbines dominate the field of both military and civil aviation; they are extensively used in air defense and naval ships, and are the major prime movers for onshore and, particularly, offshore exploration and the production of oil and gas. More recently, gas turbines have become increasingly important in both combined heat and power and in combined cycle power generation applications. Gas turbines are able to use the widest possible range of fuels from low heating value gases to coal, producing the lowest levels of pollutant emissions. Their range of application is too wide to list here, and continues to grow. The aerospace application of gas turbines has made our lives safer and richer by enabling us to safeguard our country and by creating opportunities for trade and access to the fruits of other cultures. Jet powered aircraft have provided the leading nations with unprecedented air power supremacy for defense and global reach to help promote worldwide peace and aid.

1.4 Working Principle of Aero Gas Turbine Engines

Aero gas turbine engines work on the principle of jet propulsion which is mainly the application of Newton's third law of motion, in which it is stated that "each and every force acting on a body produces an equal and opposite reaction." Basically, a gas turbine engine is a heat engine which discharges a fast moving jet of exhaust gases that generates the thrust for jet propulsion. The jet engine is regarded as an internal combustion air-breathing engine. The key component of such an engine is a gas turbine, which is used to extract energy from the engine itself to drive the compressor.

After compression in the compressor, air is heated in the combustion chamber by adding and igniting fuel, thereby releasing energy. The resulting gases are directed over the turbine blades, spinning the turbine to extract the energy, and mechanically powering the compressor. Finally, the gases are passed through a nozzle, generating additional thrust by accelerating the hot exhaust gases by expansion back to atmospheric pressure. The working cycle of a gas turbine engine is similar to that of the four-stroke piston engine. However, in a gas turbine engine combustion occurs at a constant pressure whereas, in a piston engine, it occurs at a constant volume.

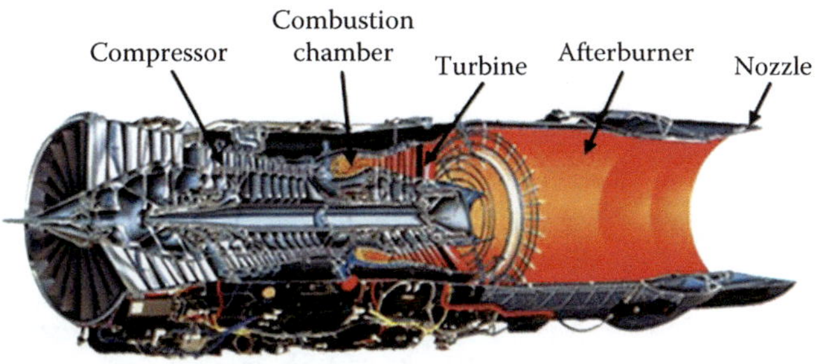

FIGURE 1.3
Critical components of an aero engine. (Courtesy of Thermonet, Department of Mechanical Engineering, University of Texas, Austin, TX, n.d., http://www.wiley.com/college/moran/CL_0471465704_S/user/tutorials/tutorial9/tut9h_content.html.)

1.5 Critical Components of an Aero Engine

The simplest form of an aero gas turbine engine consists of four main components: a compressor, a combustion chamber, a turbine, and a propelling nozzle. It is essentially a heat engine using air as working fluid to provide thrust to the aircraft. The critical components of a typical aero engine are shown in Figure 1.3.

1.6 Aero Engine Compressor

Most modern passenger and military aircraft are powered by gas turbine engines. They are also called "jet engines." All gas turbine engines have a compressor to increase the pressure of the incoming air before it enters the combustor. The performance of a compressor in aero engines has considerable influence on the total engine performance.

There are two main types of compressors: axial and centrifugal. Air flow in the axial compressor travels parallel to the axis of rotation whereas, in a centrifugal compressor, the air flow is turned perpendicular to the axis of rotation. Both types of compressor are driven by the engine turbine and are usually coupled directly to the turbine shaft. The centrifugal compressor is a single or two-stage unit using an impeller to accelerate the air and a diffuser to produce the required pressure rise. The axial flow compressor is a multistage unit using alternate rows of rotator blades and stator vanes. This arrangement in axial flow compressors (shown in Figure 1.4) accelerates and diffuses the air until the required pressure rise is obtained. A centrifugal compressor is shown in Figure 1.5. In some cases, especially in small aero engines, an axial compressor is used in combination with a centrifugal

FIGURE 1.4
An axial flow compressor. (Courtesy of Shenyang Blower Works Group, Katy, TX, n.d., http://www.sbw-turbo.com/ewebeditor/UploadFile/20151214145516634.jpg.)

FIGURE 1.5
A centrifugal compressor. (Courtesy of Concepts NREC, White River Junction, VT, n.d., http://www.conceptsnrec.com/getmedia/e3c0a482-ed58-40f6-8955-d8cc648d13a2/gallery-comp11.aspx.)

compressor (shown in Figure 1.6). In this case, the role of the axial compressor is to boost the inlet pressure to the centrifugal compressor.

With regard to the merits and demerits of the two types, a centrifugal compressor is usually more robust than an axial compressor and is also easier to develop and manufacture. However, axial flow compressors consume

FIGURE 1.6
A jet engine using combination of axial flow and centrifugal compressor. (Courtesy of Wikipedia, "Sleipnir, Lycoming T 53 gas turbine," https://en.wikipedia.org/wiki/File:Turboprop_T-53.jpg, last modified on March 6, 2017.)

more air than a centrifugal compressor of the same frontal area and can be designed to attain much higher pressure ratios. Since air flow is an important factor in determining the amount of thrust, axial flow compressors give more thrust for the same frontal area. This advantage of increasing the pressure ratio by the addition of extra stages has led to the adoption of axial compressors in most engine designs. However, centrifugal compressors, which were used in initial aero engines, are still used on small turbojets and turboshaft engines where their simplicity and ruggedness has overcome any other disadvantages.

A single spool compressor consists of one rotor assembly and stators with as many stages as necessary to achieve the desired pressure ratio, and all the airflow from the intake passes through the compressor.

A multi-spool compressor consists of two or more rotor assemblies, each driven by their own turbine at an optimum speed to achieve higher pressure ratios and to give better operating flexibility. During operation, the rotor is turned at high speed by the turbine so that air is continuously induced into the compressor, which is then accelerated by the rotating blades and swept rearwards onto the adjacent row of stator vanes. The pressure rise results from the energy imparted to the air in the rotor which increases the air velocity. The air is then decelerated (diffused) in the subsequent stator passage and the kinetic energy is converted to pressure. Stator vanes also serve to correct the direction of the air stream given by the rotor blades. The last row of stator vanes usually acts as an air straightener to remove swirl from the air prior to entry into the combustion chamber at a lesser axial velocity. The changes in pressure and velocity that occur in the airflow through the compressor are accompanied by a progressive increase in air temperature as the pressure increases.

1.7 Aero Engine Combustion Chamber

A combustion chamber must be capable of allowing fuel to burn efficiently over a wide range of operating conditions without incurring a large pressure loss. In addition, if flame extinction occurs, it must relight the gas stream. The gas turbine engine operates on a constant pressure cycle; therefore, any loss of pressure during the process of combustion must be kept to a minimum.

The temperature of the gases released by combustion is about 1800°C–2000°C, which is far too hot to be sustained by the combustor casing as well as for entry to the nozzle guide vanes of the turbine. Air not used for combustion, which amounts to 60% of the total airflow, is therefore introduced progressively into the flame tube. A part of this is used to lower the gas temperature in the dilution zone and the remainder is used for cooling the walls of the flame tube as well as the combustor casing. Combustion should be completed before the diluted air enters the flame tube, otherwise incoming air will cool the flame and incomplete combustion will result.

A gas turbine combustor must satisfy a wide range of requirements whose relative importance varies among engine types. However, the basic requirements of all combustors are listed as follows [10,11]:

- High-combustion efficiency (i.e., the fuel should be completely burned so that all its chemical energy is liberated as heat)
- Sufficient fatigue strength
- Durability
- Reliable and smooth ignition, both on the ground (especially at very low ambient temperatures) and, in the case of aircraft engines, after a flameout at high altitude
- Wide stability limits (i.e., the flame should stay alight over a wide range of pressures and air/fuel ratio)

For aircraft engines, size and weight are important considerations; for industrial engines, greater emphasis is placed on other items, such as a long operating life and multifuel capability.

1.8 Aero Engine Turbine

An aero engine turbine is employed to provide power to drive the compressor and adjuncts for engines which use the jet as the sole means of propulsion. For engines which do not use the jet for propulsion, the turbine is used to provide power to the shaft of a propeller or rotor. The turbine accomplishes this task by extracting energy from the hot gases released from the combustion system and expanding them to a lower pressure and temperature. High stresses are involved in this process and, for efficient operation, the turbine blade tips rotate at very high speeds of over 450 m/s. The continuous flow of hot gases to which the turbine is exposed leads to an entry temperature of up to 1700°C.

The basic components of the turbine are the combustion discharge nozzles, the nozzle guide vanes, the turbine discs, and the turbine blades. The rotating assembly is carried on bearings mounted on the turbine casing and the turbine shaft is common to the compressor shaft, as shown in Figure 1.7. Sometimes, the turbine shaft is connected to the compressor shaft by a self-aligning coupling.

The turbine blades are shaped like aerofoils and are considered as the most critical part in a gas turbine engine. They are subjected to the highest temperature of the hot gases and are designed in such a way to withstand the temperature. They are designed to provide passage between adjacent blades that gives a steady acceleration of the flow up to the "throat," where the area is the smallest and the velocity reaches that required at exit to produce the required degree of reaction. Tiny holes are made on the turbine trailing edge to provide cooling of the turbine during operation. The actual area of the cross-section of each blade is fixed by the permitted stress in the material used and by the size of the cooling holes. Figure 1.8 shows a turbine rotor assembled with blades.

Since turbine entry temperatures have been very high in aero engines, this has been a challenge in selecting the materials for the nozzle guide vanes

FIGURE 1.7
A jet engine with an axial flow compressor coupled with axial flow turbine. (Courtesy of Wikipedia, "General Electric, Turbojet," OH, https://en.wikipedia.org/wiki/General_Electric_J85, last modified on November 27, 2016.)

FIGURE 1.8
An axial flow turbine.

and the turbine blades. The high speed of rotation, which imparts tensile stress to the turbine disc and blades, is also a limiting factor. Nickel alloys are used in association with a cooling arrangement for the manufacture of nozzle guide vanes and blades. A ceramic coating is applied to enhance the heat resistant properties. In addition, turbine blades are manufactured either by directional solidification or from single crystals in order to have sufficient directional strength characteristics.

1.9 Aero Engine Propelling Nozzle

A propelling nozzle is the component of a jet engine that operates to constrict the flow, to form an exhaust jet that maximizes the velocity of the propelling gases from the engine. The nozzles can be subsonic, sonic, or supersonic. Convergent-divergent nozzles can give supersonic jet velocity within the divergent section, whereas in a convergent nozzle, the exhaust gases cannot exceed the speed of sound within the nozzle. Propelling nozzles can have fixed geometry or can have variable geometry. Nozzles that have variable

(a) (b)

FIGURE 1.9
(a) Converging diverging (CD) nozzle of a single engine aircraft and (b) propelling nozzles of an F-15 Eagle fighter. (Courtesy of Stack Overflow Talent Orders, New York, NY, https://www.reddit.com/r/MachinePorn/comments/320jj1/a_nozzle_on_a_mcdonnell_douglas_f15_eagle_at_the evergreen aviation and space museum in McMinville/; http://space.stackexchange.com/questions/1171/efficient-types-of-nozzles-used-in-rockets.)

geometry are designed to give different throat and exit diameters so as to deal with differences in ambient pressure, flow, and engine pressure, thus permitting improvement of thrust and efficiency.

An aero engine propelling nozzle is shown in Figure 1.9. The primary objective of a nozzle is to raise the pressure in the engine by constricting the flow at the smallest cross-section (the throat), and then to expand the stream of exhaust gases to atmospheric pressure and convert it into a high-speed jet of gases to propel the aircraft. The energy to accelerate the gases comes from the temperature and pressure of the gas. Almost all nozzles have a convergent section. Simple convergent nozzles are used on many jet engines. If the nozzle pressure ratio is above the critical value (about 1.8:1), a convergent nozzle will choke, resulting in some of the expansion to atmospheric pressure taking place downstream of the throat. In general, it is observed that narrow convergent nozzles give high-speed exhaust but reduced thrust, whereas wide convergent nozzles give a lower speed, but higher thrust. Engines capable of supersonic flight use convergent-divergent nozzles that generate supersonic flow. In a supersonic combusting ramjet (also known as a "scramjet"), the air is already at supersonic speed before entering the nozzle. Due to this, a simple divergent nozzle can be used.

Of all the major parts in an aero engine, combustors play a crucial role in determining many of an engine's operating characteristics, such as fuel efficiency, levels of emissions, and transient response (the response to changing conditions such as fuel flow and air speed). The combustor casing which houses the combustor is subjected to the highest pressure and a very high temperature during operation. Therefore, attention is focused in the design evaluation of an aero engine combustion chamber and improvement in its design.

1.10 Design Philosophy of an Aero Engine Combustion Chamber

The development of a preliminary combustion chamber design methodology and the automation of the design process would help to meet the demands of the customer. Following the recent advances of computer-aided design and computational fluid dynamics tools, some parametric design systems have already been developed within the last few years (2010–2016). Shakariyants [15] published a generic combustor design tool mainly based on geometric rules, not taking into account the methodology employed for the design validation of the combustor [16–20].

The first step in the specification of a new combustor takes place at the preliminary design stage within the development of new aircraft engines. The basis of the preliminary design is always the current state-of-the-art technology, which is characterized by appropriate design rules specific to the engine manufacturer. Limitations on the design space are provided by customer requirements, project targets, and design experience gained from previous combustor designs and tests. Based on these details and on incorporating requirements from the emission certification regulations, a new engine is characterized by a performance model, from which a compressor and turbine preliminary design is derived. The engine performance parameters, compressor exit aerodynamics, and turbine entry conditions are boundary conditions for the preliminary design of a combustor, as shown in Figure 1.10.

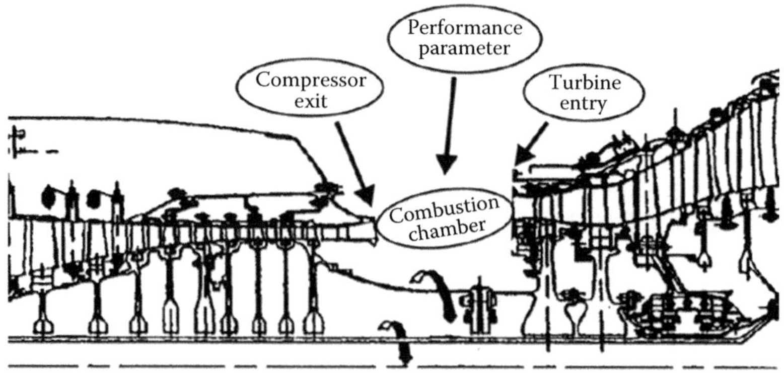

FIGURE 1.10
Initial combustor design parameters. (Data from Nima, P. and Michael, P., Development of a combustion chamber design methodology and automation of the design process, *25th International Congress of the Aeronautical Sciences*, Hamburg, Germany, pp. 1–10, 2006.)

During the initial phase of combustion chamber design, certain layout parameters are fixed. These initial parameters are the performance parameters and the geometric data for the compressor exit and the turbine entry. Based on the geometric data, geometric gauge points are derived defining the combustor entry and exit levels and areas. Assuming a design rule for the cowl entry level (e.g., compressor exit and cowl entry level are placed on the same radius), another gauge point is fixed. Based on certain empirical relationships, the volume required for efficient combustion is calculated.

In normal operation, the overall air/fuel ratio of a combustion chamber may vary between 45:1 and 130:1 [22]. However, the kerosene which is used for gas turbine fuel can burn efficiently at or close to 15:1 (known as its stoichiometric ratio). For this reason, the fuel must be burned with part of the air entering the combustion chamber in what is called the "primary combustion zone." This is achieved by means of a combustion liner that has various devices for metering the air flow distribution along the chamber. The casing is designed based on the requirements of fixed flow velocities in the inner and outer annuli, since the velocities only depend on the geometry (i.e., inner and outer annuli and height, respectively).

1.11 Types of Combustion Chamber in Aero Engines

There are three major design configurations of combustion chambers in use for gas turbine engines. They are

1. Multiple combustion chambers
2. Tubo-annular combustion chambers
3. Annular combustion chambers

1.11.1 Multiple Combustion Chambers

These types of combustion chambers are used in centrifugal compressor engines and the earlier types of axial flow compressor engines. The chambers are disposed around the engine and the compressor delivery air is directed by ducts to pass into the individual chambers. Each chamber has an inner flame tube around which there is an air casing. Air passes through the flame tube snout and also between the tube and the outer casing. The separate flame tubes are all interconnected. This allows each tube to operate at the same pressure and also allows combustion to propagate around the flame tubes during engine starting. A multiple combustion chamber is shown in Figure 1.11.

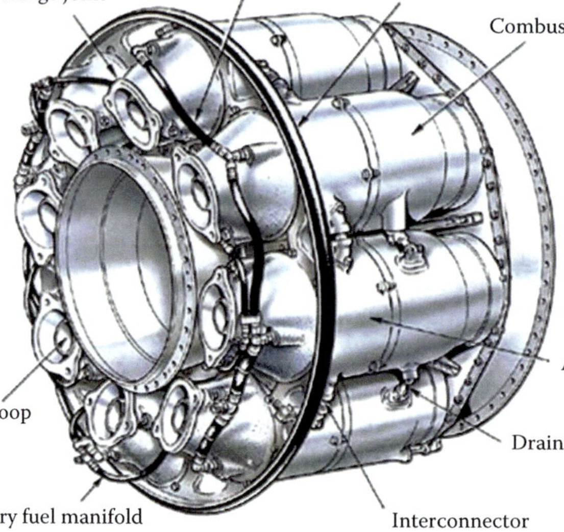

FIGURE 1.11
Multiple combustion chamber. (Data from Rolls-Royce Plc., *The Jet Engine*, 5th edn., Renault Printing Co., Birmingham, Great Britain, 1996, pp. 35–44.)

1.11.2 Tubo-Annular Combustion Chamber

The tubo-annular combustion chamber bridges the gap between the multiple and annular types. A number of flame tubes are fitted inside a common air casing. This airflow is similar to that of multiple combustion chambers. This arrangement combines the ease of overhaul and testing of multiple systems with the compactness of the annular system. A tubo-annular combustion chamber is shown in Figure 1.12.

1.11.3 Annular Combustion Chamber

This type of combustion chamber consists of a single flame tube, completely annular in form, which is contained in inner and outer casings. The chamber is open at the front to the compressor and at the rear to the turbine nozzles. The main advantage of the annular chamber is that, for the same power output, the length of the chamber is only 75% of that of a tubo-annular system of the same diameter, resulting in a considerable saving in weight and production cost. Another advantage is the elimination of combustion propagation problems from chamber to chamber.

In this type of chamber, an annular liner is mounted concentrically inside an annular casing. In many ways, it is the ideal form of chamber as its clean aerodynamic layout results in a compact unit of lower pressure loss than other combustor types. Its main drawback stems from the heavy buckling

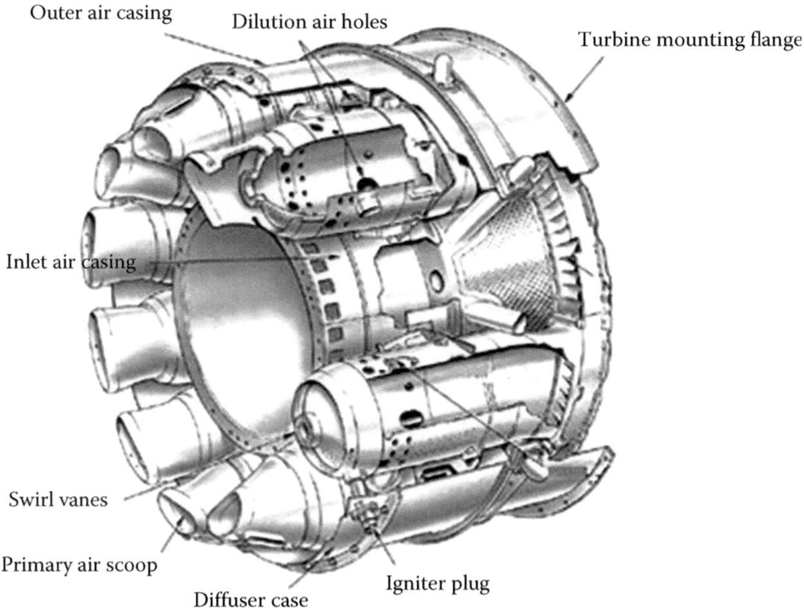

FIGURE 1.12
Tubo-annular combustion chamber. (Data from Rolls-Royce Plc., *The Jet Engine*, 5th edn., Renault Printing Co., Birmingham, Great Britain, 1996, pp. 35–44.)

load on the outer liner. Thus, in the early days of turbojet development, the use of annular liners was confined to engines with a low pressure ratio. Another drawback is the very high cost of supplying air at the levels of pressure, temperature, and flow rate required to test large annular combustion chambers in full-load conditions. Figure 1.13 shows the configuration of the annular combustors in service today.

1.12 Complexities in Combustion Chamber Design

The combustor casing is subjected to high stresses and temperatures, with a consequent risk of failure leading to the catastrophic failure of the entire engine. The consequences of failure in general are classified in various levels. A small loss of thrust, particularly if it is gradual, does not carry a high risk factor. An abrupt change or a total loss of thrust during takeoff is hazardous, although it may not be so at a higher altitude. The highest risk category of failures involves the release of high energy debris that could damage the aircraft, either structurally or functionally. The first step in the process of establishing and minimizing the risk of failure of aircraft engines is to perform

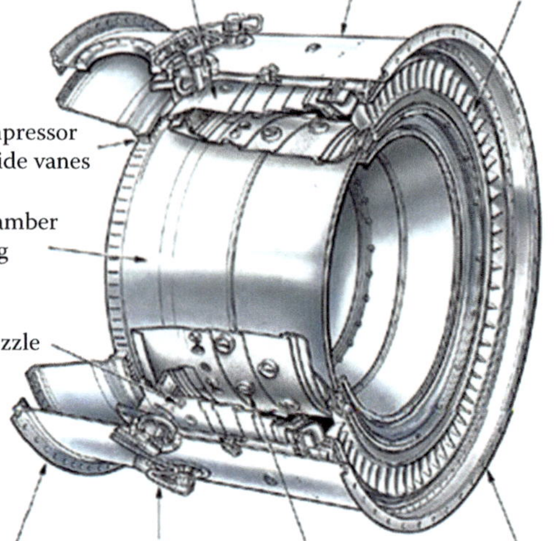

FIGURE 1.13
Annular combustion chamber. (Data from Rolls-Royce Plc., *The Jet Engine*, 5th edn., Renault Printing Co., Birmingham, England, 1996, pp. 35–44.)

a failure mode and effect analysis to identify the possible consequences of failure of an individual part, and of multiple failures in service.

The analysis undertaken in this study resulted in classification of aero engine components into three categories, each of which are expected to meet the relevant airworthiness requirement [23]:

- *Category I Component*: Highly stressed or critical parts, the failure of which might lead to engine failure, loss of engine power, or otherwise jeopardize the safety of the air craft. Example: compressor, turbine, combustor, nozzle.
- *Category II Component*: Highly stressed parts not covered in category I. Failure of these component leads to subsequent failure of the class I components. Example: pipelines, gear box, fuel system, bearing housing, lubrication system.
- *Category III Component*: Lightly stressed parts, which have secondary functions and failure of which would not seriously affect safety or performance of the engine, meaning that the engine can still be run without catastrophic failure, for example, noncritical small parts.

The combustor casing is classified as a class I component, as it experiences the highest pressure in the aero engine and its failure can have catastrophic effects that result in the destruction of the engine. Therefore, special attention should be given in designing the combustor casing to ensure it meets life expectancy as well as safety requirements of the engine. To encounter the stringent strength and life requirements, the components should meet the design strength requirements. The strength and life requirements are specified in MIL-E-5007 D standards with respect to duty cycles specified for the engine components, and the design strength requirements for compressors, combustors, turbines, shafts are the criteria for selection of materials [24].

1.13 Materials Used for Combustion Chamber

The containing walls and internal parts of the combustion chamber must be capable of resisting the high gas temperature in the primary zone. In practice, this is achieved by using the best heat resistant materials available, by the use of high heat resistant coatings and by cooling the inner wall of the flame tube as insulation from the flame.

The combustion chamber must also withstand corrosion due to the products of the combustion, high temperature, and fatigue due to fatigue stresses. Present gas turbine engines use nimonic super ally C263 having material oxidation capability up to 900°C. It is mainly made of nickel (52%), cobalt (20%), chromium (20%), molybdenum (6%), titanium (2.4%), iron (0.7%), manganese (0.6%), silicon (0.4%), copper (0.2%), and carbon (0.06%).

This combustor is housed inside an annular casing called combustor casing and this casing is generally made out of the heat resistant super alloy Inconel 718 [25–27]. This is a precipitation hardened nickel-based super alloy and is typically used in high temperature applications and has the following advantages:

- High strength at working temperature
- Better strength-to-weight ratio
- Resistance to oxidation and corrosion
- Ease of fabrication and weldability compared with other materials of a similar category
- Better coefficient of linear expansion
- Easy availability

Inconel 718 is a relatively recent alloy; its industrial use started in 1965. It contains significant amounts of iron (Fe), niobium (Nb), and molybdenum (Mo). Minor contents of aluminum (Al) and titanium (Ti) are also present. Nickel (Ni) and chromium (Cr) contribute to the corrosion resistance of this material. Inconel 718 combines good corrosion and high mechanical properties with an excellent weldability. It is employed in modern gas turbines, rocket engines, turbine blades, and in extrusion dies and containers [28,29]. A problem with nickel alloys is their high density of 8–9 g/cm^3, which is about twice as dense as titanium and three times denser than aluminum. Lighter materials are used in aero engine components whenever possible to reduce the engine weight. Titanium alloys are used to reduce the engine weight, but their use is restricted to components in the fan and compressor sections where the temperature is less than 550°C. Titanium is used on the leading edges of carbon-fiber fan blades.

In this research work, an attempt is made to evaluate the fatigue life of an aero engine combustor casing made out of Inconel 718 alloy to meet the mandatory airworthiness requirements. Significant contributions made in this research work are outlined here.

1.14 Significant Contributions

- Development of a state-of-the-art servo hydraulic pressure test facility working in a closed loop to evaluate the fatigue life of a combustor casing to meet the airworthiness requirement
- Design and manufacture of the test setup for adapting the combustor casing in the test facility
- Identification of the critical locations and the mounting of strain gauges at those locations
- Carrying out the test in simulated engine conditions
- Analysis of the critical zone and correlation of the experimentally obtained result with the analysis data
- Recommendations for changes in the geometric features of the component and changes in the manufacturing process to afford improved fatigue life
- Manufacturing of a modified combustor casing in accordance with the method recommended
- Instrumentation of the modified combustor casing, and testing and evaluation of its fatigue life
- Safety test of the modified combustor casing
- Studying the effect of fatigue on the overpressure capability of the combustor casing
- Ensuring the application of the combustor casing in a combat aero engine

1.15 Summary

This chapter presented the development details in the field of aero engines, critical aero engine parts, and aero engine combustion chambers. The technical literature in the field of fatigue design and life evaluation methods is presented in detail in Chapter 2.

2

Fatigue Design Philosophy of an Aero Engine Combustor Casing

2.1 Introduction

This chapter presents a summary of the literature reviewed during the course of this study and publication. The review of the literature is expected to provide the relevant background information, to identify the gaps, and to highlight the state of technology available to date and, thus, to formulate the objectives of the present investigation. The primary objective of the present work is to evolve a comprehensive experimental methodology for the life evaluation of a gas turbine combustor under cyclically varying pressure loads. Designing a combustor to meet aero gas turbine requirements has proven to be a highly difficult task since the evolution of the gas turbine. Various research works have been carried out related to gas turbine combustor design, manufacturing methodology, lifing, and post-failure analysis; and these are reflected in the ensuing paragraphs.

2.2 Combustion Chamber Design

By the 1960s, the annular layout of the combustor casing was firmly established as the automatic choice for all new aircraft engines. From this period, and throughout the 1980s, the most important annular combustors were those fitted to the GE CF6, P&W JT9D, and RR RB211 engines. These engines were all highly successful, both technically and commercially. Improvements in wide-body aircraft, along with continuing market pressures to reduce cost, called for engines in the 355–450 kN thrust class with growth potential up to around 500 kN. To meet this demand, Rolls Royce developed the Trent engine, which is a direct descendant of the RB211 series. The GE90 and the P&W4084 both achieve similar performance and thrust levels. All three engines are fitted with annular combustors, which embody the latest advances in fuel injection and wall-cooling techniques.

Modern gas turbine combustors are made of high-temperature alloys, employ effusion cooling, and are protected by a thermal barrier coating (TBC). Gas turbine combustor failure modes—such as TBC spalling, cracking, and distortion resulting from oxidation, creep, and thermal fatigue—are driven by hot-spot peak temperature and the associated thermal gradient. Standard material characterization tests—such as creep, oxidation, and low-cycle fatigue (LCF)—are indicators of a material's potential performance but they fully represent neither the combustor geometric/material system nor the thermal fatigue conditions a combustor is subjected to during engine operation. Combustor rig tests and/or engine cyclic endurance tests to determine the suitability of new material systems for combustors are time-consuming and costly.

Koff [30,31] emphasized the durability of the combustor liner and its life improvement in his research work during 2004. However, the combustor casing which houses the liner is not considered in his work.

Rudrapatna [32] developed an efficient test method for screening material systems under representative combustor conditions. He has developed an experimental system in his work to fill the gap. The work was focused on configured specimen geometry, test methodology, observed test results, and a comparison with typical failure modes observed in combustors.

The combustor casing which is considered for the present study is fabricated by forming/machining and, subsequently, the adaptors/parts are joined by welding. There have been many approaches suggested for the estimation of the fatigue life of welded joints. The majority of work published in the scientific literature is concerned with predicting fatigue in welded joints of simple geometry, such as can conveniently be made into test specimens. Taylor and Hoey [33] developed a crack modeling method for the prediction of fatigue failures in welded joints. However, this model was found to give large conservative errors. During predictions, potential problems have occurred when the component dimensions are small with respect to a critical dimension. Some solutions have been developed for particular cases but more work needed in this area.

The fatigue strength of welded structures in high-strength materials depends mainly on the local weld geometry and the magnitude of residual stress after welding [34–36]. Barsoum [37] evaluated the fatigue strength of high-strength steel joints by considering the residual stresses in and around the heat-affected zone of the weld metal in the welding process. The residual stresses are static and are imposed on the cyclic stresses during fatigue loading [38,39].

Failure of welded structures leads to various direct losses, such as the cost of repair work and the cost of work to prevent future failure as well as accident compensation [40]. Sehgal [41,42] developed a model to identify the stage relationships among various cause events and to evaluate the importance of each cause event for the failure modes. This model provides direction for appropriate action for the minimization of a failure mode, thus improving the reliability of the joints.

Tao et al. [43,44] were instrumental in retrieving a failed aero engine from an air crash in 2000 and investigated the cause of the failure. It was observed that the accident was caused by the failure of the combustor liner. One of the regions in the liner was torn open and had a long axial crack along one welding seam. They carried out metallographic analysis, mechanical calculation, and fatigue life estimation of the combustor to determine the nature of and reason for the failure. The failure was due to fatigue cracking in poor quality roll welding. This occurrence emphasizes the importance of welding in the successful running of an aero engine combustor.

2.3 Fatigue Failure in Aero Engines

Designers should focus on the various failure modes due to which the combustor casing may fail during its complex mission loading. It has been seen by the researchers that, in the case of critical aero engine components, the most predominant mode of failure is fatigue.

Fatigue is a type of fracture that occurs in materials that are subjected to changing or varying stresses over time. Fatigue is a major problem in many of the modern machines and devices we use today, such as airplanes, automobiles, locomotives, bridges, and many mechanical machines. Fatigue is mainly caused by the environment in which the materials are utilized.

A failure in an aircraft jet engine can have severe consequences which cannot be accepted and high requirements are therefore imposed on engine reliability. Consequently, assessment of the reliability of life predictions used in design and maintenance are important.

Gas turbine components experience severe cyclic thermal-mechanical loading in the high-pressure compressor, turbine modules, and the combustion chamber. Therefore, it is a requirement of the gas turbine designer to study the fatigue behavior of the materials used for these modules in detail. The oldest and still most commonly used method for estimating the fatigue life of metallic components is the S-N curve, even though the drawbacks are well known. One of them is that the fatigue life of notched specimens differs considerably from those observed for smooth specimens. The local strain method has been assumed to overcome this problem. In this method, it is suggested that the same strain amplitude always leads to the same fatigue life, regardless of the specimen shape. Evan et al. [45] developed a test procedure to test Inconel specimens to establish the thermo-mechanical fatigue characteristics of a material. The experiment was conducted on cylindrical specimens. The procedure highlighted the overview of the test procedure.

Since the combustor casing is made of Inconel alloy in sheet form, an attempt is made to extract the mechanical properties from the specimens in sheet form as an input to the numerical analysis work carried out as a part of this book.

Fatigue is the result of the cumulative process consisting of crack initiation, propagation, and final rupture of a component. It is the progressive, localized, and permanent structural change that occurs in a material subjected to repeated or fluctuating strains at nominal stresses that have maximum values less than (and often much less than) the static yield strength of the material. Fatigue may culminate in cracks and cause fracturing after a sufficient number of fluctuations. Fatigue damage is caused by the simultaneous action of cyclic stress, tensile stress, and plastic strain. If any one of these three is not present, a fatigue crack will not initiate and propagate. The plastic strain resulting from cyclic stress initiates the crack; the tensile stress promotes crack growth (propagation). Although compressive stresses will not cause fatigue, compressive loads may result in local tensile stresses. Microscopic plastic strains also can be present at low levels of stress where the strain might otherwise appear to be totally elastic.

During fatigue failure in a metal that is free of crack-like flaws, microcracks form, coalesce, or grow to macrocracks that propagate until the fracture toughness of the material is exceeded and finally fracturing occurs. Under usual loading conditions, fatigue cracks initiate near or at singularities that lie on or just below the surface, such as scratches, sharp changes in cross-section, pits, inclusions, or embrittled grain boundaries. Microcracks may be initially present due to welding, heat treatment, or mechanical forming. Even in a flaw-free metal with a highly polished surface and no stress concentrators, a fatigue crack may form. If the alternating stress amplitude is high enough, plastic deformation (i.e., long-range dislocation motion) takes place, leading to slip steps on the surface. Continued cycling leads to the initiation of one or more fatigue cracks. Alternately, the dislocations may pile up against an obstacle, such as an inclusion or grain boundary, and form a slip band, a cracked particle, decohesion between particle and matrix, or decohesion along the grain boundary.

The initial cracks are very small. Their size is not fully known because it is difficult to determine when a slip band or other deformation feature becomes a crack. Certainly, however, cracks as small as a fraction of a micron can be observed using modern metallographic tools such as the scanning electron microscope or scanning tunneling microscope. The microcracks then grow or link up to form one or more macrocracks which, in turn, grow until the fracture toughness is exceeded.

During cyclic loading, localized plastic deformation occurs at the highest stress site. This plastic deformation induces permanent damage to the component and a crack develops.

As the component experiences an increasing number of loading cycles, the length of the crack (damage) increases. After a certain number of cycles, the crack will cause the component to fail (separate). Cracks start on the localized shear plane or at near high stress concentrations, such as persistent slip bands, inclusions, porosity, and discontinuities.

The localized shear plane usually occurs at the surface or within grain boundaries. Crack nucleation is the first step in the fatigue process.

Once nucleation has occurred and cyclic loading continues, the crack tends to grow along the plane of maximum shear stress and through the grain boundary.

Figure 2.1 presents a graphical representation of the fatigue damage process, showing where crack nucleation starts at the highest stress concentration site(s) in the persistent slip bands.

The next step in the fatigue process is the crack growth stage. This stage is divided between the growth of Stage I and Stage II cracks. Stage I crack nucleation and growth are usually considered to be the initial short crack propagation across a finite length of the order of a couple of grains on the local maximum shear stress plane. In this stage, the crack tip plasticity is greatly affected by the slip characteristics, grain size, orientation, and stress level, because the crack size is comparable to the material microstructure.

Stage II crack growth refers to long crack propagation normal to the principal tensile stress plane globally and in the maximum shear stress direction locally. In this stage, the characteristics of the long crack are less affected by the properties of the microstructure than the Stage I crack. This is because the crack tip plastic zone for a Stage II crack is much larger than the material microstructure.

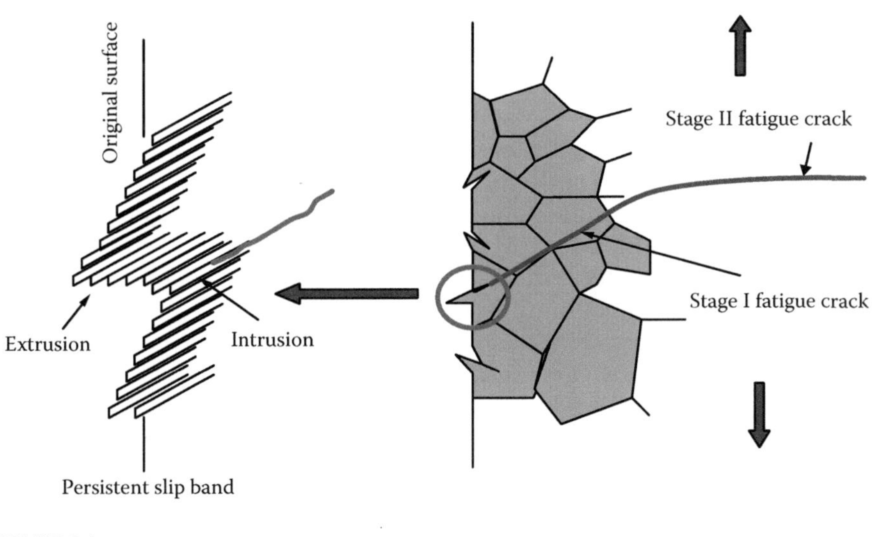

FIGURE 2.1
The fatigue process: a thin plate under cyclic tensile loading [52]. (Data from Yung, L.-L. et al., *Fatigue Testing and Analysis: Theory and Practice*, Elsevier Butterworth, Oxford, U.K., 2005, pp. 57–76.)

In engineering applications, the amount of component life spent on crack nucleation and short crack growth is usually called the "crack initiation period," whereas the component life spent during long crack growth is called the "crack propagation period."

Therefore, in general, the total fatigue life of a component can be divided into three phases: crack initiation, propagation, and final rupture of the component. The last phase is rapid and can be ignored in practical work, while calculating the total fatigue life. Stable crack growth can be reliably modeled using linear elastic fracture mechanics. Large numbers of formulae for stress intensity factors for different geometries and loading cases can be found in the literature today and the use of weight functions offers a possibility to expand the use of the method.

Typically, the crack initiation period accounts for most of the fatigue life of a component made of steels, particularly in the high-cycle fatigue (HCF) regime (approximately > 100,000 cycles).

In the LCF regime (approximately < 100,000 cycles), most of the fatigue life is spent on crack propagation. Once a crack has formed or complete failure has occurred, the surface of a fatigue failure can be inspected. A bending or axial fatigue failure generally leaves behind beach markings. The name of these markings comes from the appearance of the surface. The crack nucleation site is the center of the shell, and the crack appears to propagate away from the nucleation site, usually in a radial manner. Within the beach lines are the striations. These striations represent the extension of the crack during one loading cycle. These rings appear similar to the rings on the cross-section of a tree. Instead of one ring for each year of growth, there is one ring for each loading cycle. In the event of a failure, there is a final shear lip, which is the last bit of the material supporting the load just before failure. The size of this lip depends on the type of loading, the material, and other conditions.

LCF has been described as a progressive failure phenomenon brought about by the cyclic application of strains that extend to plastic range to produce failure in about 10^5 cycles or less. LCF is characterized by cyclic loads, which are a relatively high significant plastic strain and a low number of cycles to failure [46,47]. Many of the aircraft engine components, including the diffuser casing, experience LCF and, hence, it is necessary to study the lifing methodology in brief for these components [48–51].

Therefore, much research has been done in the field of fatigue fracture resistance to counter the effect of varying stress on the material.

Even recently, as reported during 2011, Southwest Flight 812 suffered from a catastrophic fatigue fracture which happened 18 minutes into its flight. The fuselage skin ripped open, causing a massive loss in cabin pressure, and the aircraft was forced to land on a remote military base in Yuma, Arizona. The cause of the damage is not always apparent due to the nature of fatigue fracture often occurring at low stresses and elevated temperatures in most materials. Accidents such as this have been a leading motivation for fatigue research.

Investigators revealed after inspection that tiny cracks had been found on the Flight 812 aircraft, around rivet holes along a joint where two areas of the fuselage skin overlap. They were trying to determine how long the cracks had been there. An inspection found "persistent fatigue along the entire fracture surface" [56].

Goswamy [53–55] developed an interactive mechanism of different processes, such as fatigue, creep, and so on, which produce premature failures in engineering components used in gas turbines, propulsion engines, structures, and other systems and subsystems. Corran and Williams [56–58] describe the regulatory requirements that must be achieved to allow operation of aero engines in civil applications and examine how these requirements are satisfied in practice.

The evaluation of fatigue life starts with a good approximation of the mission profile or a duty cycle of the aircraft. The stress amplitude with respect to time may be expected to vary in a typical mission profile, often referred to as "spectrum loading." This spectrum loading condition is difficult to simulate in the testing environment. A most widely used method, called the "rainflow counting technique," is used to convert the complex mission profile to a simple maximum–minimum cycle.

Cycles can be counted using time histories of the loading parameter of interest, such as force, torque, stress, strain, acceleration, or deflection.

The objective of this section is to introduce several cycle counting techniques that can be used to reduce a complicated variable amplitude loading history into a number of discrete, simple, constant, amplitude loading events, which are associated with fatigue damage. In fatigue, damage is usually interpreted as crack growth. It is well known that fatigue crack growth depends more on the effective stress or strain range than on the applied ranges. The effective values result from the crack closure phenomenon due to which the applied stress range is reduced; such a mechanism strongly depends on the load history.

2.4 Fatigue Cycle Counting Methods

Cycle counting is used to summarize lengthy, irregular load-versus-time histories by providing the number of times cycles of various amplitudes occur. The definition of a cycle varies with the method of cycle counting. In fatigue analysis, a cycle is the load variation from the minimum to the maximum and then to the minimum load. Cycle counts can be made for time histories of force, stress, strain, torque, acceleration, deflection, or other loading parameters of interest.

Various counting methods are applied to extract amplitude and number of thermal cycles from a mission profile. Unfortunately, the estimated lifetime may change, depending on the method used. To discover the correct counting technique for lifetime calculation, the finite element method was used to simulate the thermomechanical stresses experienced by the power module when it is submitted to a given thermal load.

A proper interpretation of a complex load, stress, or strain versus time histories requires that an approximate cycle counting method must be used. There are eight cycle counting methods, as described as follows:

1. *Peak count method*: This method identifies the occurrence of a relative maximum or minimum load value. The peak counting method has been widely used since it is easy to count cycles manually. In this method, all maximums above the mean and all minimums below the mean are counted. A count is recorded each time the history shows a change in slope. Peaks and valleys are ignored if the signal has not recrossed the reference. Counts are combined to form the largest possible cycle first. Small amplitude loadings are eliminated by mean-crossing peak counting method. The most damaging cycle count for fatigue analysis is derived from the peak count by first constructing the largest possible cycle, using the highest peak and lowest valley, followed by the second largest cycle, etc., until all peak counts are used. This method is the most selective one considered as only the maximum excursion from the datum is counted between successive datum crossings.
2. *Peak between mean-crossing count method*: This method may be called mean-crossing peak count method or maximum peaks between zero crossing method. In this method, only the largest peak between successive crossings of the mean is counted, which is equivalent to completely neglecting the stress variations between two passages through zero. The signal is thus reduced to only one peak between two passages through zero. This type of counting can lead to incorrect results.
3. *Level-crossing count method*: As the name suggests, the method counts every crossing of the predetermined stress levels. In this method, all positive slope level crossings above the mean and negative slope level crossings below the mean are counted. Reference load crossings are counted on the positive sloped portion of the loading history. Restrictions on the level-crossing counts are often specified to eliminate small amplitude variations which can give rise to a large number of counts. To do this, filter small load excursions prior to cycle counting. Or, make no counts at the reference load and specify that only one count be made between successive crossings of a secondary lower level associated with each level above the reference load. The counting of level crossings neglects small load variations when they are in one class. This is not the case, however, when a

class limit is crossed (by however small an amount). For this reason the result of this counting method depends on the magnitude of the amplitude suppression. The result of the counting is a cumulative frequency which is usually presented in a spectrum. This method is used to count continuous load-time histories for subsequent analysis of fatigue. The significant advantage of this counting method is that it is one-parametric and can easily be put in a graphical form in a so called spectrum. The disadvantage is that for some load-time histories, especially with mean stress variations, this counting method gives too severe a spectrum.

4. *Fatigue-meter count method*: In this method, one count is made between successive crossings of a lower level associated with each counting level.
5. *Range count method*: This counting technique defines one count as a range, the height between a successive peak and valley. Positive ranges and negative ranges are defined on positively sloped reversals and negatively sloped reversals, respectively. Each range represents on one-half cycle (reversal). In this method, each range, that is, the difference between successive peak values is counted as 1/2 cycle, the amplitude of which is half the range.
6. *Range-mean count method*: In this method, ranges are counted as mentioned earlier, and the mean value of each range is also considered. This method analyzes fluctuations in both stress amplitude and mean stress. The signal is divided into half cycles (valley-peak or peak-valley) and their stress amplitude and mean are recorded. This method has the potential of underestimating the severity of the high amplitude signal, while the almost negligible damage due to small cycles is included. This method is not recommended, since it may miss out large excursions that are detrimental to fatigue life.
7. *Range pair method*: This method counts a strain range as a cycle if it can be paired with a subsequent straining of equal magnitude in opposite direction. For a complicated history, some of the ranges counted as cycles will be simple ranges during which the strain does not change direction, but others will be interrupted by smaller ranges that will also be counted as cycles.
8. *Rainflow method*: This method of counting the cycles is more widely used than any other method. The strain time history (or stress time history) is plotted so that the time axis is vertically downward, and the lines connecting the strain peaks are imagined to be a series of sloping roofs. The rainflow method is a method for counting fatigue cycles from a time history. The fatigue cycles are stress-reversals. The rainflow method allows the application of Miner's rule in order to assess the fatigue life of a structure subject to complex loading.

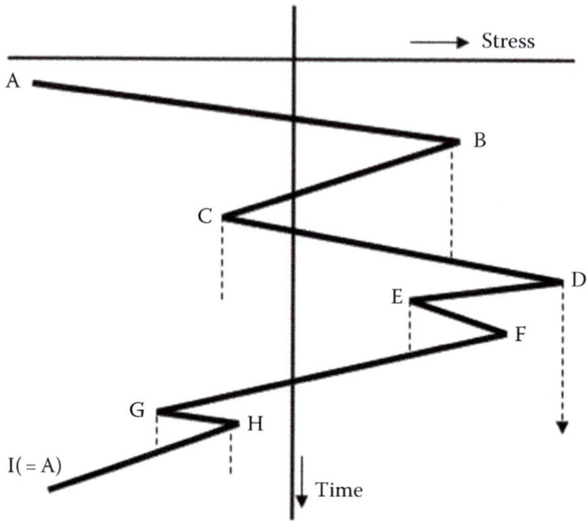

FIGURE 2.2
Rainflow counting method.

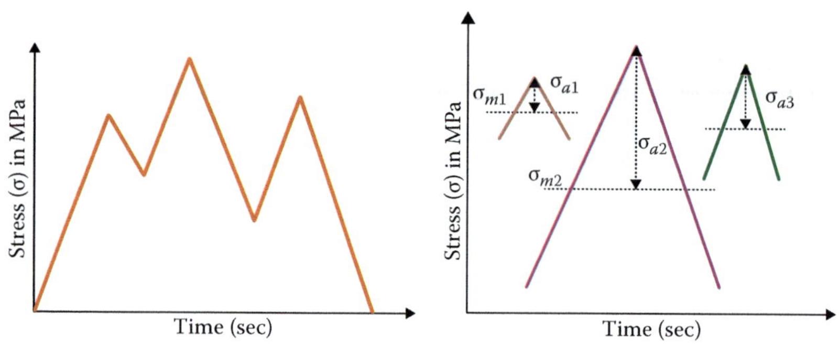

FIGURE 2.3
A simple example of the rainflow counting method.

This counting method is most widely used by most of the researchers working in the field of fatigue life evaluation of aero engine components. This method of cycle counting is shown in Figures 2.2 and 2.3, and is explained in detail in the following section.

The rule for counting the cycle is as follows:

- The rain is allowed to flow on the roof and dip down to the next slope except that, if it initiates a valley, it will terminate when it meets a valley more negative than the valley from which it initiated.

- Similarly, if the rainflow initiates a peak, it will terminate when it meets a peak more positive than the peak from which it initiated.
- A rainflow must also stop if it meets the rain from a roof above.
- If cycles are to be counted over the duration of a duty cycle or a mission profile block that is to be repeated block after block, the cycle counting should be started by initiating the first rain drop either at the most negative valley or the most positive peak, and continuing until all cycles in a complete block have been counted in sequence.

In aerospace industries, highest priority is given to the safety of employees, customers, and the airplane. The aero engines that are used for aircraft propulsion must meet certain airworthiness requirements.

Cruse and Meyer [59] developed an LCF damage model primarily meant for the life extension of gas turbine engine rotors. They used specimens, subcomponents and full-scale aero engine rotors to develop and substantiate the proposed model.

Biswas et al. [60,61] brought out the role of fatigue in the failure of turbine blades in an air turbine starter, giving insight as to the need for further studies and experimental evaluations of the effect of fatigue on aero engine components.

Since the component is subjected to high pressure and high temperatures in the engine, all aero engine manufacturers use nickel-based heat resistant alloys for the fabrication of combustor casings.

Sunder and Nelson [62–64] carried out extensive research analysis and testing to examine fatigue crack and propagation on materials under specially designed complex load sequences. They observed that cycle counting is an important element of fatigue analysis and it is reasonable to expect that cycle counting retains its significance in the analysis of the fatigue mechanism [65–68].

The importance of fatigue crack growth analysis to ensure the reliability of critical components under cyclic loading conditions is shown in the research work carried out by Singh et al. [69,70]. The work was intended to evaluate accurately the fatigue life of structures/components having multiple discontinuities such as holes, cracks, and/or inclusions.

It has been experimentally proved that fatigue behavior seems to be relatively insensitive to a change in waveform and cyclic rate as long as the peak values are same. Hence, a sinusoidal waveform as well as ramp loading (with the same peak stress) essentially produce the same result as far as fatigue life is concerned. A ramp waveform of constant amplitude and time period is usually considered for fatigue loading during fatigue life estimation in test rigs [71–74].

Jacobsson et al. [75] carried out experiments on Inconel 718 specimens subjected to fatigue loading and the fractured surface is shown in Figure 2.4. The fracture surfaces showed a dominance of trans-granular crack propagation with striations, indicating a low degree of time dependency in the procedure [76,77].

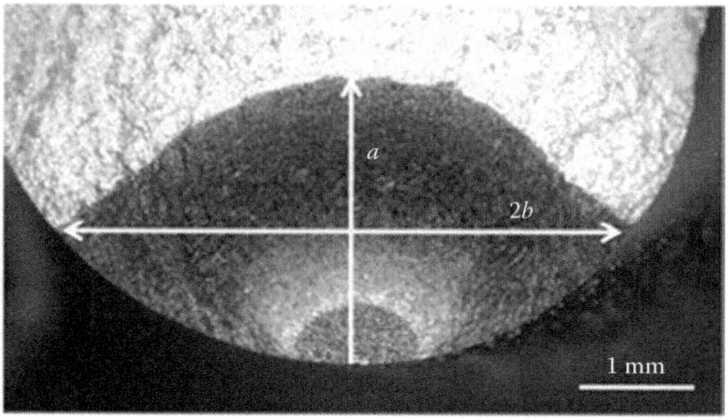

FIGURE 2.4
Morphology of a fatigue crack observed on an Inconel 718 specimen.

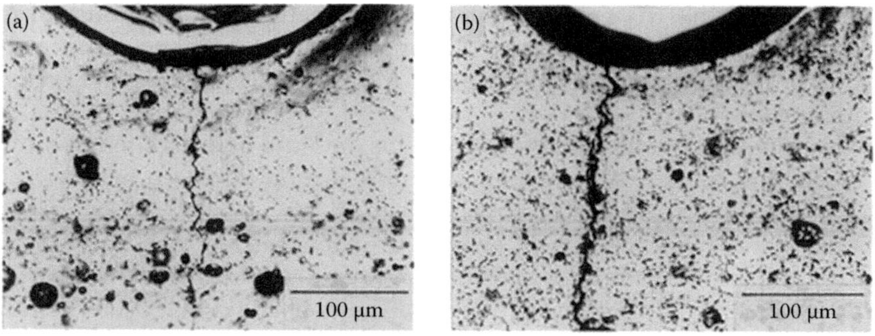

FIGURE 2.5
Optical micrographs showing the morphology of a crack initiated from the rim of the holed specimen: (a) at room temperature and (b) at 500°C.

The fatigue crack initiates only on the surface and then propagates to the body of the component causing the final failure. Experiments carried out on Inconel 718 specimens at room temperature as well as at an elevated temperature demonstrate crack initiation and propagation, as illustrated in Figure 2.5 [78–80].

2.5 Fatigue Life Evaluation Methods

Predicting fatigue damage for structural components subjected to variable loading conditions is a complex issue. The first, simplest, and most widely used damage model is linear damage. The rule is often referred to as Miner's rule.

A component operating in spectrum loading conditions undergoes fatigue damage for a given stress amplitude and number of cycles. The complex mission profile is converted to simple minimum–maximum cycles using the rainflow counting technique. Each cycle is responsible for producing a certain amount of fatigue damage to the component. This fatigue damage is cumulative in nature, according to the Miner's hypothesis.

Operation of spectrum loading over different stress levels results in a damage fraction D_i for each of the different stress levels σ_i in the spectrum. When these damage fractions sum up to unity, failure is predicted.

Failure is predicted to occur if:

$$D_1 + D_2 + D_3 + \cdots + D_{i-1} + D_i > 1 \tag{2.1}$$

where $D_i = n_i/N_i$; that is, the damage factor (D_i).

At any level, σ_i is linearly proportional to the ratio of number of cycles of operation (n_i) to the total number of cycles (N_i) that would produce failure at that stress level.

Failure is predicted to occur if:

$$\frac{\Sigma n_j}{N_j} > 1 \tag{2.2}$$

If this sum is less than unity, then the amount of cumulative damage occurred by all the stress cycles is computed and, from this, the number of cycles required to cause the same amount of damage by a single minimum–maximum stress cycle is obtained.

This theory is widely used because of its simplicity. However, in many cases the linear rule often leads to nonconservative life predictions. The results from this approach do not take into account the effect of load sequence on the accumulation of damage due to cyclic fatigue loading.

Miner's rule applies if the cyclic load can be approximated by the number of cycles times an average amount of load per cycle of a certain size. However, if the load is some nonlinear function of the number of cycles, then Miner's rule is invalid.

Since the introduction of the linear damage rule, many different fatigue damage theories have been proposed to improve the accuracy of fatigue life prediction [81–84]. In certain cases, Miner's rule is modified to obtain the exact behavior of the material or the component. Sun [85] made an attempt to revise Miner's rule that eliminates a drawback in the original which distinguishes between a component population and an individual population, and takes into account that the damage accumulation to the point of failure for different individuals in the population is not a constant but, rather, a random variable. The formulae for calculating the equivalent loads are then derived for different variable loads. Requirements for the correct use of these formulae are also proposed. The equivalent load of a type of practical component is calculated and verified by test data. The result shows that the revised Miner's rule is more reasonable and the derived formulae possess a higher level of precision that can be applied to practical engineering.

Nian et al. [86] were instrumental in developing a model using Miner's rule to predict the life of marine and offshore structures. According to Yao [87,88], fatigue damage increases with the applied loading cycles in a cumulative manner and the material deteriorates with the corrosion time. A cumulative fatigue damage rule under the alternative of corrosion or cyclic loading was proposed. The specimens of aluminum alloy LY12-CZ soaked in corrosive liquid for different times were tested under constant amplitude cyclic loading to obtain S-N curves. The test was carried out to verify the proposed cumulative fatigue damage rule under different combinations of corrosion time loading level and the cycle numbers. It was shown that the predicted residual fatigue lives accorded well with the experimental results and that the proposed rule was simple and could be easily adopted.

Seam welding technology is widely employed to join two or more structural parts together in a variety of industries, including ship construction and the offshore, pressure vessel and piping, automotive, and aerospace industries. Generally, a seam weld can be created by various welding techniques including resistance seam welding, gas tungsten arc welding (GTAW), gas metal arc welding (GMAW), shielded metal arc welding (SMAW), submerged arc welding (SAW), and laser welding. Kang and Lee [89,90] discussed fatigue life prediction methods of welded joints through their work on different seam welded joints. Welding strongly affects the materials by the process of heating and subsequent cooling, as well as by the fusion process with additional filler material, resulting in inhomogeneous and different materials. Furthermore, a weld is usually far from perfect, containing inclusions, pores, cavities, undercuts, and so on. The shape of the weld profile and nonwelded root gaps creates high stress concentrations with varying geometry parameters. Moreover, residual stresses and distortions due to the welding process affect fatigue behavior. Nominal stress approaches are relatively simple but have limitations when it comes to disclosing stresses and strains at the critical regions at the welded joint.

Where the stress is high enough for plastic deformation to occur, the account in terms of stress is less useful and the strain in the material offers a simpler description.

In case of complex geometries involving a weld, analysis is generally carried out without the weld. The stress thus obtained is multiplied by a factor called the "weld factor." For simplification and to obtain a swift result, it is a general practice to carry out linear analysis and incorporate Neuber's rule to obtain equivalent nonlinear stress and strain values. Neuber's methodology [91–95] is most widely used in evaluating the fatigue lives of critical aero engine components. Neuber's rule is depicted in Figure 2.6.

Karakas [96] assessed the fatigue strength of three types of weld joints based on the mean stress damage parameter model. These analyses are performed

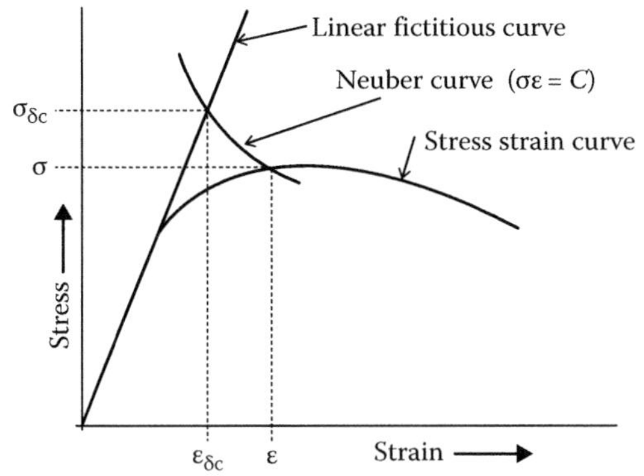

FIGURE 2.6
Neuber's rule.

with three different stress ratios, $R = -1$, $R = 0$ and $R = 0.5$, under axial loading. For each stress level, the corresponding Neuber hyperbolas, together with their maximum stress and maximum strain values, are determined in order to calculate damage parameter values. For a given weld geometry, this damage parameter is able to unify the fatigue results for different R-values within at a tight scatter band and therefore to consider the mean stress effect.

LCF is usually characterized by the Coffin–Manson relation (published independently by Coffin in 1954 and Manson in 1953) [97,98]. Thus, the damage occurring in an LCF operation can be related to plastic strain amplitude by the Coffin–Mansion equation, which is explained in the following.

2.6 Fatigue Damage Assessment

It has been more than 50 years since the development of the LCF law. The Coffin–Manson relationship, sometimes also referred to as the Manson–Coffin relationship, is proposed independently by Stan Manson and Coffin. Most fatigue studies limit their focus on material behavior under a single well-defined set of loading conditions. It is assumed that the material is characterized under a wide range of such loading conditions. How the material behaves under a sequence of different loadings has come to be known as "cumulative fatigue," even though a single loading also involves the cumulative effects of several fatigue processes. Many cyclic loadings involved in a service involve complex conditions characterized by a hysteresis loop which has a mean stress. The width of the hysteresis loop implies plastic flow which should result in a finite life. That is essentially the nature of the Coffin–Manson equation.

As pointed out by Manson and Coffin, the use of total strain range in fatigue life equations has the advantage that it does not require computation of relative amounts of elastic and plastic strain ranges.

As fatigue properties are often not available by experimentation, many suggestions were proposed to obtain the first approximation estimates from simple tensile data, such as the Universal Slopes equation of Manson and Coffin [99–101]. The total strain in a component under LCF conditions is given in Equation 2.3 as follows:

$$\Delta\varepsilon_t = \Delta\varepsilon_p + \Delta\varepsilon_e \tag{2.3}$$

$$\Delta\varepsilon_t = D(N_f)^{\gamma} + C(N_f)^{\beta} \tag{2.4}$$

where

$\Delta\varepsilon_t$ is the total strain
$\Delta\varepsilon_p$ is the plastic component of the strain
$\Delta\varepsilon_e$ is the elastic component of the strain
D is the ductility coefficient
C is the strength coefficient
N_f is the number of cycles to failure
γ is the fatigue ductility exponent
β is the fatigue strength exponent

The coefficients D, C, γ, and β are determined from the experiments.

When the loading cycle is a completely reversed cycle, the alternating stress amplitude is computed as half the stress range in the cycle, having the mean stress equal to zero. The amount of damage caused by a stress cycle depends not only on the alternating stress, but also on the mean stress. Two different cycles having the same alternating stress but different mean stresses cause different amounts of damage. Therefore, the mean stress applied on a component must be considered while evaluating fatigue life.

Changing the mean stress tends only to have effects in relation to HCF. The effect of mean stress is obliterated by LCF problems due to the higher plasticity. Therefore, the mean stress is accounted for only on the elastic side of the equation. The modified Coffin–Manson equation [102] suggested by Manson and Halford is illustrated in Equation 2.5:

$$\nabla\varepsilon_t = D^a N_f^b + \frac{3.5}{E}\{\text{UTS} - \sigma_{\text{mean}}\} N_f^b \tag{2.5}$$

$$\nabla\varepsilon_t = D^{0.6} N_f^{-0.6} + \frac{3.5}{E}\{\text{UTS} - \sigma_{\text{mean}}\} N_f^{-0.12} \tag{2.6}$$

in which elastic and plastic lines are taken with constant slopes for all materials. Fatigue strength is correlated to ultimate tensile strength (UTS) and fatigue strain is approximated by the ductility $D = -\ln(1 - \text{RA})$ in a monotonic tensile test calculated by a specimen reduction area, where

E is Youngs' modulus of the component

UTS is the ultimate tensile strength of the material

a, b are the fatigue exponents and are evaluated from experiments

If the fatigue properties are known, a life prediction—and, hence, a damage factor and an LCF life—can be computed from the earlier equation. The fatigue life thus obtained must be verified by experimental techniques. Many research works have been carried out in the field of validation techniques wherein the numerical techniques are validated with another numerical/analytical technique or by experimentation. Experimental techniques are found as the best methodology for validation. Experiments are conducted with specimens at most academic institutions; these offer economical results because it is cost-effective to prepare specimens and prepare a test setup. These experiments only give approximate results compared with a full-scale test result.

To identify the exact behavior of a component, structural tests need to be conducted on a full-scale component by applying loads in accordance with the operating environment. Full-scale testing on aero engine components takes time and money. Considerable effort is needed to manufacture the test component and to set up a test facility and testing arrangement. For aero engine houses, it is a mandatory requirement to carry out experiments on full-scale test components for certification and design clearance of the aero engine components. The component is inspected before start of the test as well as at a regular interval during the test by nondestructive testing (NDT) methods, as explained in the following section.

2.7 Nondestructive Testing Methods

NDT comprises a wide group of analysis techniques used in science and industry to evaluate the properties of a material, component, or system without causing damage to the component. The terms nondestructive examination (NDE), nondestructive inspection (NDI), and nondestructive evaluation (NDE) are also commonly used to describe this technology. Because NDT does not permanently alter the article being inspected and does not destroy the integrity and serviceability of the part or system, it is a highly valuable technique that can save both money and time in product evaluation, troubleshooting, and research. In other words, when the inspection or test is completed, the part can still be used.

In contrast to NDT, other tests are destructive in nature and are therefore undertaken on a limited number of samples ("lot sampling"), rather than on the materials, components, or assemblies actually being put into service. These destructive tests are often used to determine the physical properties of materials, such as impact resistance, ductility, yield and ultimate tensile strength, fracture toughness, and fatigue strength, but discontinuities and differences in material characteristics are more effectively found by NDT.

Today, modern nondestructive tests are used in manufacturing, fabrication, and in-service inspections to ensure product integrity and reliability, to control manufacturing processes, to lower production costs, and to maintain a uniform quality. During construction, NDT is used to ensure the quality of materials and joining processes during the fabrication and erection phases, and in-service NDT inspections are used to ensure that the products in use continue to have the integrity necessary to ensure their usefulness and the safety of the public.

While being a high technology concept, evolution of the equipment has made it robust enough for application in any industrial environment at any stage of manufacture—from steel-making to site inspection of components already in service. A certain degree of skill is required to apply the techniques properly in order to obtain the maximum amount of information concerning the product, with consequent feedback to the production facility.

NDT is not just a method for rejecting substandard material; it is also an assurance that the supposedly good is good. The technique uses a variety of principles; there is no single method around which a black box may be built to satisfy all requirements in all circumstances.

A brief description of the methods [103] most commonly used in industry, together with details of typical applications, functions, and advantages, are as follows:

- Radiography (x-ray) inspection
- Magnetic particle inspection (MPI)
- Dye penetrant inspection (DPI)
- Ultrasonic inspection
- Eddy current and electromagnetic inspection

2.7.1 Radiography Inspection

Industrial radiography involves exposing a test object to penetrating radiation so that the radiation passes through the object being inspected and a recording medium placed against the opposite side of that object. For thinner or less dense materials such as aluminum, electrically generated x-radiation (x-rays) is commonly used; for thicker or denser materials, gamma radiation is generally used.

Gamma radiation is given off by decaying radioactive materials, with the two most commonly used sources of gamma radiation being Iridium-192 (Ir-192) and Cobalt-60 (Co-60). IR-192 is generally used for steel up to 2½–3 in., depending on the Curie strength of the source, and Co-60 is usually used for thicker materials due to its greater penetrating ability.

The recording media can be industrial x-ray film or one of several types of digital radiation detectors. With both, the radiation passing through the test object and exposes the media, causing darker areas where more radiation has passed through the part and lighter areas where less radiation has penetrated. As the isotope is continuously emitting radiation, it must be housed in a container of depleted uranium or a similar dense shielding material, so as to protect the environment and personnel.

To produce an x or gamma radiograph, the film package (comprising film and intensifying screens—the latter being required to reduce the exposure time—enclosed in a light proof cassette) is placed close to the surface of the subject. The source of radiation is positioned on the other side of the subject some distance away, so that the radiation passes through the subject and onto the film. After the exposure period, the film is removed, processed, dried, and then viewed by transmitted light on a special viewer. Various radiographic and photographic accessories are necessary, including such items as radiation monitors, film markers, image quality indicators, darkroom equipment, and so on. Regarding developing the film, there are many degrees of sophistication, including fully automatic processing units. These accessories are the same for both x and gamma radiography systems. Other consumable items, such as as radiographic film and processing chemicals, are also required.

If there is a void or defect in the part, more radiation passes through, causing a darker image on the film or detector, as shown in Figure 2.7.

- *Advantages of radiography inspection*:
  - Information is presented pictorially.
  - A permanent record is provided which may be viewed at a time and place distant from the test.
  - Useful for thin sections.
  - Sensitivity declared on each film.
  - Suitable for any material.
- *Disadvantages of radiography inspection*:
  - Generally, an inability to cope with thick sections.
  - Possible health hazard.
  - Need to direct the beam accurately for two-dimensional defects.
  - Film processing and viewing facilities are necessary, as is an exposure compound.

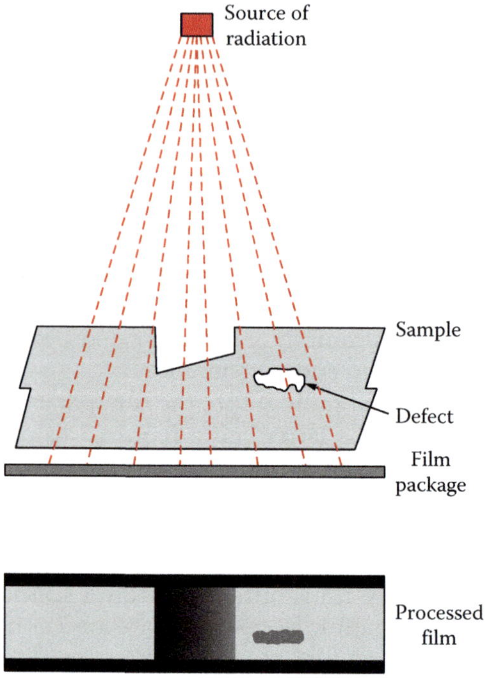

FIGURE 2.7
Radiography inspection technique.

- Not suitable for automation, unless the system incorporates fluoroscopy with an image intensifier or other electronic aids.
- Not suitable for surface defects.
- No indication of the depth of a defect below the surface.

2.7.2 Magnetic Particle Inspection

MPI uses one or more magnetic fields to locate surface and near surface discontinuities in ferromagnetic materials. The principle is to generate magnetic flux in the article to be examined, with the flux lines running along the surface at right angles to the suspected defect. Where the flux lines approach a discontinuity, they will stray out into the air at the mouth of the crack. The crack edge becomes magnetic attractive poles, North and South. These have the power to attract fine particles of magnetic material, such as iron filings. Usually these particles are of an oxide of iron in the size range 20–30 μm and are suspended in a liquid which provides mobility for the particles on the surface of the test piece, assisting their migration to the crack edges. However, in some instances they can be applied in a dry powder form. The particles can be red or black oxide, or they can be coated with a substance which fluoresces brilliantly under ultraviolet illumination (black light). The object is to present as great a contrast as possible between the indication of the crack and the material background.

The technique not only detects those defects which are not normally visible to the unaided eye, but also renders easily visible those defects which would otherwise require close scrutiny of the surface. There are many methods with which to generate magnetic flux in a test piece, the most simple is the application of a permanent magnet to the surface, but this method cannot be controlled accurately due to indifferent surface contact and deterioration in magnetic strength. Modern equipment generates the magnetic field electrically, either directly or indirectly.

The magnetic field can be applied with a permanent magnet or an electromagnet. When using an electromagnet, the field is present only when current is being applied. When the magnetic field encounters a discontinuity transverse to the direction of the magnetic field, the flux lines produce a magnetic flux leakage field of their own. Because magnetic flux lines do not travel well in air, when very fine colored ferromagnetic particles ("magnetic particles") are applied to the surface of a test piece, the particles will be drawn into the discontinuity, reducing the air gap and producing a visible indication on the surface of the part. The magnetic particles may be a dry powder or suspended in a liquid solution, and they may be colored with a visible dye or a dye that fluoresces under an ultraviolet ("black") light. The magnetic particle inspection technique is illustrated in Figure 2.8.

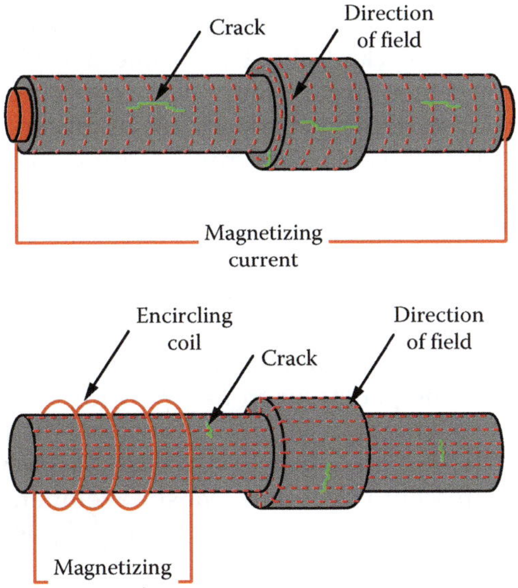

FIGURE 2.8
Magnetic particle inspection technique.

- *Advantages of magnetic particle inspection*:
    - Simplicity of operation and application.
    - Quantitative.
    - Can be automated, apart from viewing. (However, modern developments in automatic defect recognition can be used in parts with a simple geometry; for example, billets and bars. In this case, a special camera captures the defect indication image and processes it for further display and action.)
- *Disadvantages of magnetic particle inspection*:
    - Restricted to ferromagnetic materials.
    - Restricted to surface or near surface flaws.
    - Not fail safe, in that a lack of indication could mean either no defects or that the process was not carried out properly.

2.7.3 Dye Penetrant Inspection

This method is also known as "liquid penetrant inspection." The basic principle of DPI is that when a very low viscosity (highly fluid) liquid (the penetrant) is applied to the surface of a part, it will penetrate into fissures and voids open to the surface. Once the excess penetrant is removed, the penetrant trapped in those voids will flow back out, creating an indication. Penetrant testing can be performed on magnetic and nonmagnetic materials, but does not work well on porous materials. Penetrants may be "visible," meaning they can be seen in ambient light, or under fluorescence, requiring the use of a "black" light. The visible dye penetrant process is shown in Figure 2.9. When performing a DPI inspection, it is imperative that the surface being tested is clean and free of any foreign materials or liquids that might block the penetrant from entering voids or fissures open to the surface of the test piece. After applying the penetrant, it is permitted to sit on the surface for a specified period of time (the "penetrant dwell time"), then the part is carefully cleaned to remove excess penetrant from the surface. When removing the penetrant, the operator must be careful not to remove any penetrant that has flowed into voids. A light coating of developer is then be applied to the surface and given time ("developer dwell time") to allow the penetrant from any voids or fissures to seep up into the developer, creating a visible indication. Following the prescribed developer dwell time, the test piece is inspected visually, with the aid of a black light for fluorescent penetrants. Most developers are fine-grained, white talcum-like powders that provide a color contrast to the penetrant being used.

The process is purely mechanical/chemical and the various substances used may be applied in a large variety of ways, from aerosol spray cans at

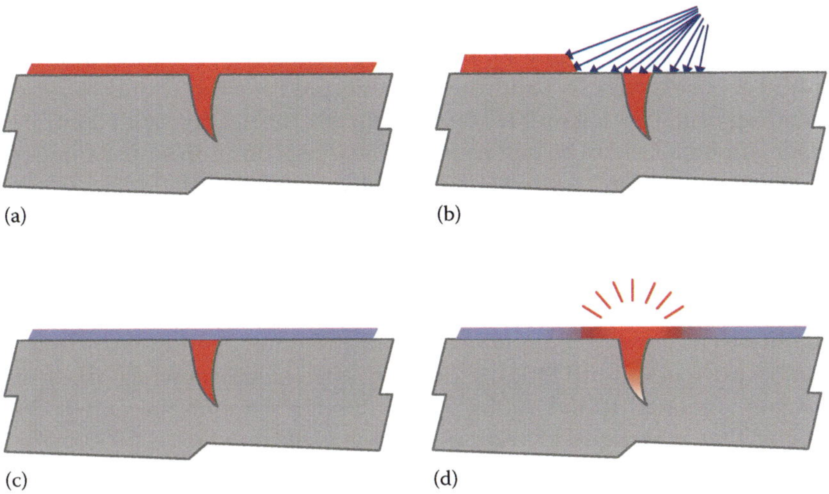

FIGURE 2.9
Dye penetrant inspection technique: (a) penetrant applied to the surface and enters defect, (b) excess penetrant removed from surface, (c) developer powder applied to draw penetrant out of crack, and (d) accentuated indication of crack as penetrant spreads around the opening.

the most simple end to dipping in large tanks on an automatic basis at the other. The latter system requires sophisticated tanks and spraying and drying equipment but the principle remains the same.

- *Advantages of dye penetrant inspection*:
    - Simplicity of operation
    - Best method for surface breaking cracks in non-ferrous metals
    - Suitable for automatic testing, with reservations concerning viewing (See automatic defect recognition in advantages of magnetic particle inspection in Section 2.7.2)
    - Quantitative
- *Disadvantages of dye penetrant inspection*:
    - Restricted to surface breaking defects only
    - Decreased sensitivity
    - Uses a considerable amount of consumables

2.7.4 Ultrasonic Inspection

Ultrasonic testing uses the same principle as is used in naval sonar. Ultrahigh-frequency sound is introduced into the part being inspected and, if the sound hits a material with a different acoustic impedance (density and acoustic velocity), some of the sound will reflect back to the sending unit and can be presented on a visual display. By knowing the speed of the sound through the part (the acoustic velocity) and the time required for the sound to return to the sending unit, the distance to the reflector (the indication with the different acoustic impedance) can be determined. The most common sound frequencies used in ultrasonic transduction (UT) are between 1.0 and 10.0 MHz, which are too high to be heard and do not travel through air. The lower frequencies have greater penetrating power but less sensitivity (the ability to "see" small indications), while the higher frequencies do not penetrate as deeply but can detect smaller indications. Vibrations or sound waves at 1–10 MHz frequency have the ability to travel a considerable distance in homogeneous elastic material, such as many metals with little attenuation. The velocity at which these waves propagate is related to the Young's modulus for the material and is characteristic of that material. For example, the velocity in steel is 5900 m/s, and in water 1400 m/s. Ultrasonic energy is considerably attenuated in air, and a beam propagated through a solid will, on reaching an interface (e.g., a defect, or intended hole, or the backwall) between that material and air reflects a considerable amount of energy in the direction equal to the angle of incidence.

The two most commonly used types of sound wave used in industrial inspections are the compression (longitudinal) wave and the shear (transverse) wave, as shown in Figure 2.10. Compression waves cause the atoms in a test piece to vibrate back and forth parallel to the direction of the sound and shear waves cause the atoms to vibrate perpendicularly (from side to side)

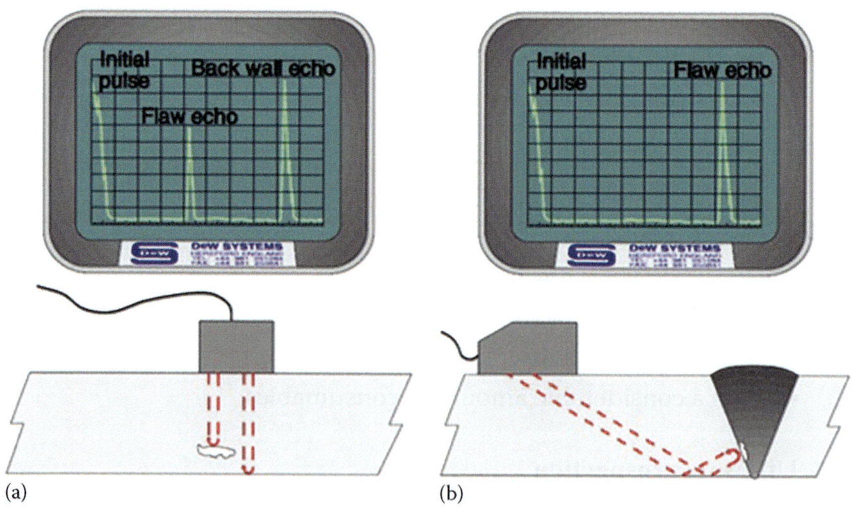

FIGURE 2.10
Ultrasound inspection technique: (a) schematic diagram of ultrasonic detection of slag in steel section using a normal probe and (b) schematic diagram of the use of an angle probe to detect defects not directly under the probe. Such as in weld inspection.

to the direction of the sound. Shear waves travel at approximately half the speed of longitudinal waves.

Sound is introduced into the test piece using an ultrasonic transducer ("probe") that converts electrical impulses from the UT machine into sound waves, then converts returning sound back into electric impulses that can be displayed as a visual representation on a digital or liquid crystal display (LCD) screen (on older machines, a cathode ray tube [CRT] screen). If the machine is properly calibrated, the operator can determine the distance from the transducer to the reflector and, in many cases, an experienced operator can determine the type of discontinuity (such as slag, porosity, or cracks in a weld). Because ultrasound will not travel through air (the atoms in air molecules are too far apart to transmit ultrasound), a liquid or gel called a "couplant" is used between the face of the transducer and the surface of the part to allow the sound to be transmitted into the part. This technique is used for the detection of internal and surface (particularly distant surface) defects in sound conducting materials.

- *Advantages of ultrasonic inspection*:
    - Thickness and lengths up to 30 ft can be tested.
    - Position, size, and type of defect can be determined.
    - Instant test results.
    - Portable.
    - Extremely sensitive if required.
    - Capable of being fully automated.
    - Access to only one side necessary.
    - No consumables.
- *Disadvantages of ultrasonic inspection*:
    - No permanent record available unless one of the more sophisticated test results and data collection systems is used.
    - The operator can decide whether the test piece is defective or not while the test is in progress.
    - Indications require interpretation (except for digital wall thickness gauges).
    - Considerable degree of skill necessary to obtain the fullest information from the test.
    - Very thin sections can prove difficult.

2.7.5 Eddy Current and Electromagnetic Inspection

Eddy current testing uses the fact that, when an alternating current coil induces an electromagnetic field into a conductive test piece, a small current is created around the magnetic flux field, much like a magnetic field is generated around an electric current. Eddy currents can be produced in any electrically conducting material that is subjected to an alternating magnetic field (typically 10 Hz–10 MHz). The alternating magnetic field is normally generated by passing an alternating current through a coil. The coil can have many shapes and can comprise between 10 and 500 turns of wire. The magnitude of the eddy currents generated in the product is dependent on conductivity, permeability, and the setup geometry. Any change in the material or geometry can be detected by the excitation coil as a change in the coil impedance. The most simple coil comprises a ferrite rod with several turns of wire wound at one end; and is positioned close to the surface of the product to be tested. When a crack, for example, occurs in the surface of the test piece, the eddy currents must travel further around the crack and this is detected by a change in the impedance. The flow pattern of this secondary current, called an "eddy" current, will be affected when it encounters a discontinuity in the test piece, and the change in the eddy current density can be detected and used to characterize the discontinuity causing that change. A simplified schematic of eddy currents generated by an alternating current coil ("probe") is shown in Figure 2.11. By varying the type of coil, this test method can be applied to flat surfaces or tubular products. This technique works best on smooth surfaces and has limited penetration, usually less than ¼ in.

Encircling coils are used to test tubular and bar-shaped products. The tube or bar can be fed through the coil at a relatively high speed, allowing the full cross-section of the test object to be interrogated. However, due to the direction of the flux lines, circumferentially oriented discontinuities may not be detected with this application.

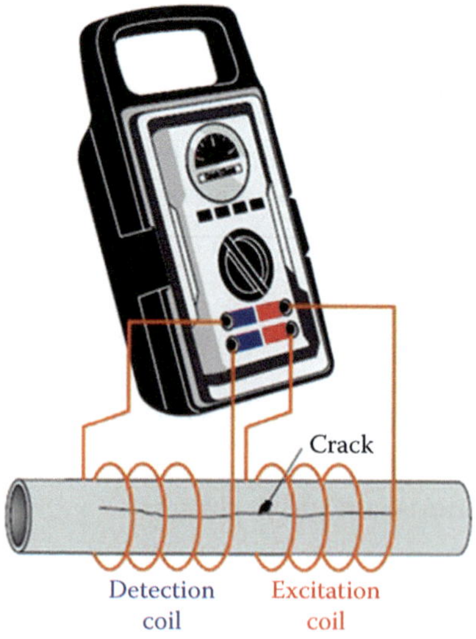

FIGURE 2.11
Eddy current inspection technique.

The main applications of the eddy current technique are for the detection of surface or subsurface flaws, conductivity measurement, and coating thickness measurement. The technique is sensitive to the conductivity of a material, its permeability, and the dimensions of a product.

- *Advantages of eddy current and electromagnetic inspection*:
  - Suitable for the determination of a wide range of conditions of conducting material, such as defect detection, composition, hardness, conductivity, permeability, etc., in a wide variety of engineering metals.
  - Information can be provided in simple terms: often go/no go. Phase display electronic units can be used to obtain much greater product information.
  - Extremely compact and portable units are available.
  - No consumables (except probes—which can sometimes be repaired).
  - Flexibility in selection of probes and test frequencies to suit different applications.
  - Suitable for total automation.
- *Disadvantages of eddy current and electromagnetic inspection*:
  - The wide range of parameters which affect the eddy current responses means that the signal from a desired material characteristic (e.g., a crack) can be masked by an unwanted parameter (e.g., a change in hardness). Careful selection of probe and electronics will be needed in some applications.
  - Generally tests restricted to surface breaking conditions and slightly subsurface flaws.

This section gave an introduction to various NDT techniques to address different testing techniques that rely on several physical phenomena while addressing the wide range of its applications. This was done in an effort to highlight several types of nondestructive evaluations and their ability to accommodate multitudes of fields and tests.

2.8 Summary of the Design Philosophy

From the earlier-mentioned philosophy, it is concluded that, in most of cases, researchers have carried out works related to the fatigue life evaluation of structures and machine components, as well as machinery. Many have developed analytical models for predicting the fatigue life of critical components that are subjected to varying loads or varying temperature. Work has also been carried out on the investigation of failure of components in service. Post-fracture investigations are carried out to find the cause of the failure and attempts are made to prevent that type of failure in future.

Ascribed to the published works in the literature, it is also observed that systematic studies on the subject of numerical analyses coupled with prediction methodologies for the onset of fatigue life and fatigue damage, and the propagation behavior of such damage at critical locations in the different types of aero engine components, have been very limited because of the complexities of geometry and complex loading conditions, and behavior of materials in different temperature conditions. The effects of stress raisers on damage evolution should not be ignored for a thorough understanding of fracture and failure phenomena in a critical aero engine component. An aero engine combustor is a very critical and classified component, and there are very limited published articles available in the field of fatigue life evaluation and its validation for the combustor. The structure of the combustor is very complex due to the constraints at the compressor exit and turbine entry, and the instrumentation requirements of the engine. It is a mandatory requirement of the certifying authorities to validate the design of critical aero engine components before they are declared fit to fly. Therefore, an attempt has been made to evaluate the fatigue life of an aero engine combustor to validate its design. The aim also has been set to put forth suitable recommendations for the improved performance of the combustor casing under consideration.

2.9 Important Design Considerations for Combustor Casing

- Experimental determination of fatigue life of an aero engine combustor casing by developing a state-of-the-art test facility and test setup.
- Identification of critical parameters by comparing the experimentally determined fatigue life with respect to the designed fatigue life of the combustor casing.
- Design and manufacturing modifications of the combustor casing for improved fatigue life.
- Validation of the modified combustor casing by subjecting it to cyclic pressure test and proof pressure test.
- Study the effect of fatigue on proof strength of an aero engine combustor casing.
- Certification and acceptance of the modified combustor casing in accordance with airworthiness requirements.

2.10 Summary

This chapter provided:

- An exhaustive review of research works on aero engine combustor design, failure investigation, and the related technologies.
- The knowledge gap in earlier investigations.
- The objectives of present research.

To meet the airworthiness requirements of an aero engine component, the LCF life of an aero engine combustor needed to be experimentally demonstrated in simulated rig tests. The engine operating conditions were considered when designing a suitable fatigue test facility and a test facility was developed. The details of the experimental procedure are explained in Chapter 3.

3

Development of Test Facility and Test Setup

3.1 Introduction

The increasing need for lightweight design, combined with improved functionality, quality, and, economic build, continues to drive the need for comprehensive testing within key structural systems and components in the aerospace sector.

Casings of aircraft engines are designed to withstand high loads from multiple directions and large numbers of fatigue cycles throughout their life. To validate their design, the fatigue life and the maximum load capability must be ensured before they are introduced into service. To simulate the environments and load conditions, specifically designed test facilities are developed and the respective components are mechanically tested for their structural integrity. Structural integrity is defined as the capability of an engine component or a system to perform satisfactorily under the operating environments for its specified life. The testing process typically includes the development of the test concept, all necessary computer simulations, the design and adaptation of the facility, mechanical equipment and adjacent dummy parts, and the proper instrumentation of the specimens (e.g., with strain gauges). This type of test usually is capable of running for several months in 24/7 testing mode using fully automatic controls.

"Mechanical testing" is a general term which refers to a broad range of activities involved with the determination and evaluation of mechanical properties and the behavior of material, structural components, and machines [104]. It is the process of applying forces, pressures, displacements, torque, or heat to a component or mechanical system and then measuring its response. The objective is to characterize the behavior of the component or the system, either in order to obtain assurance of the performance, or to verify its structural integrity. Mechanical testing is generally carried out to verify the overall performance of a mechanical system or a structural member. On the other hand, mechanical testing is used to verify the appropriateness or accuracy of a particular concept or theory. In this case, the testing is usually conducted on a simplified component or model that may actually bear no outward resemblance to a practical design. Sometimes, mechanical tests are carried out to characterize the mechanical properties of a new material.

The test methodology for all these cases is either deterministic or probabilistic in nature. In the first case, the approach is to carry out thorough and extensive tests with a single or, at most, a very few components which are assumed to be representative of the actual component. Sometimes, tests are conducted on the actual component to validate its design. This type of tests involves more cost and an expensive test setup. The second approach is to use a large number of nominally identical specimens which are assumed to constitute a suitable subset or sample of the larger set of all possible specimens. Statistical methods are then used to analyze the data and to draw conclusions about the properties of the whole set.

With an increasing demand for high thrust and light weight, the development of aero engines is a very challenging task. The design is very elaborate and stringent, as it involves high component loads (both steady and cyclic), susceptibility to multiple failure modes, and complex mission profiles. Major loads on the combustor casing are the pressure and temperature. The combustor casing designed for aero engines for military applications must meet certain stipulated strength and life requirements in accordance with military standards and specific requirements regarding the structural integrity of the engine. During the design stage, the recommended practice is to check the ability of the combustor casing to withstand the pressure and thermal load without any deformation or crack and, effectively, to meet the fatigue life and safety margins to avoid catastrophic failure of the engine during the entire flight envelope of the engine.

To evaluate the fatigue life of the combustor casing, analytical methodology is followed. However, this methodology needs to be validated by full-scale testing of the component. Testing the component in the engine carries the risk of damaging the entire engine in the event that the component fails during testing. Therefore, it is common practice to test the component in a suitably designed test facility by simulating the engine conditions in the rig [105–110]. The test also gives input with regard to the following objectives:

- To validate the design of the component
- To evaluate low cycle fatigue life and weld integrity through experiment
- To validate the lifing methodology
- To certify the component for airworthiness
- To evaluate the operating stresses in the combustor casing under cyclic pressure conditions and to arrive at casing strength margins
- To obtain strain distribution in the diffuser casing for comparison and correlation with analysis data
- Design improvements for better fatigue life

3.2 Airworthiness and Certification

Airworthiness can be defined as "fitness to fly." It is the demonstrated capability of the aircraft to perform safely within the prescribed limits. It can be further elaborated as a definition to mean the continued capability of the aircraft/engine to perform satisfactorily and fulfill its mission requirements throughout its specified life in the prevailing environments with acceptable levels of safety and reliability.

Certification is the procedure which provides the possibility of making certain that any aircraft, whether civil or military, has an acceptable safety level for a given future use within the defined flight load spectrum for a specific period. This certificate is issued by an authorized certification agency. It provides a guarantee and safety assurance to use an aircraft/systems/subsystems for the customer (services).

Certification takes place following the evaluation of a design for airworthiness. Consideration of airworthiness begins in the initial stages; namely, preliminary design, detailed design and ground tests, flight tests, and design conformance/compliance statements. The approaches/procedures differ from manned aircraft, engines for unmanned aerial vehicles, parachutes, Aircraft General Standard parts, electronic equipment, and so on. It includes all aircraft, helicopters, engines, and systems, as well as items of equipment, material, and so on, used in the development/production of aircraft, aero engines, systems, equipment, armament stores, flight suits, and so on.

The major considerations are

- Flight safety
- Assured life
- Redundancy management
- Deterioration due to life cycle environment

The different certification approaches followed are

- Continued airworthiness
- License manufacture
- System upgrades
- Ab initio development projects

It is a mandatory requirement to conduct life and safety tests on full-scale components of an aero engine for their airworthiness certification. The following section describes the details of the test facility developed for conducting life and safety tests on an aero engine combustor casing.

3.3 Description of the Test Facility and Its Subsystems

A pressure test facility was developed to carry out the test on the diffuser casing. Pressurizing the component with air, and its control, requires more sophisticated equipment, as air is compressible. It also carries the risk of a sudden release of energy when a component develops a crack, causing failure that could be catastrophic and dangerous to human life. Therefore, it was decided to use hydraulic fluid to pressurize the component. The specifications of the oil used for pressurization are as follows:

Trade name: Servo system 32

Kinematic viscosity: 29–33 cSt at 40°C

Viscosity index: 95 (minimum)

Flash point: 190°C

Pour point: −6°C

This oil is antiwear hydraulic oil specially formulated for the efficient lubrication of a wide variety of industrial equipment. This oil is blended from highly refined base stocks and carefully selected antioxidant, antiwear, antirust, and antifoam additives. It has the following benefits:

- It provides good wear protection, thereby prolonging the service life of moving parts.
- It possesses high film strength properties, which help minimize friction and wear.
- It provides rust-protection, to reduce the incidence of corrosion.
- It has excellent resistance to oxidation, thereby giving a long service life.
- It has a reduced tendency to foam, thereby minimizing the chance of noisy operation and erratic pump response.

The schematic of the facility is given in Figure 3.1.

The major subsystems of the test facility are as follows:

1. Hydraulic power supply (HPS)
2. Servo valve with manifold
3. Safety relief valve
4. Servo controller
5. Pressure transducer

The facility is designed to have two independent channels with a maximum pressure rating of 200 bar. Two different independent pressure cycles and

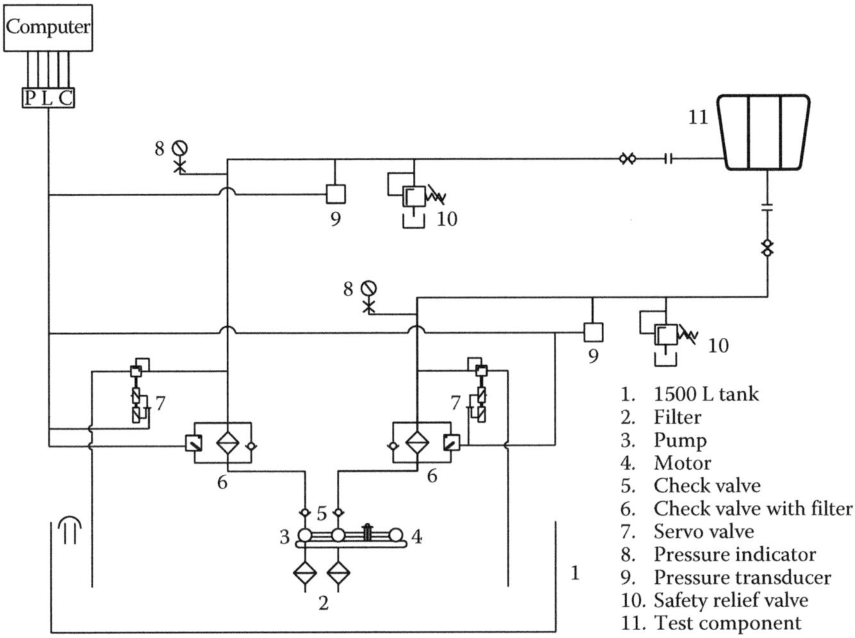

FIGURE 3.1
Schematic of the pressure test facility.

profiles can be executed in these two channels. The details of the major subsystems are described in the next section.

3.3.1 Hydraulic Power Supply

A hydraulic power pack powered by a 10 HP electric motor is used to drive a twin pump mounted in series to provide pressure to two different zones.

Flow rate: 5.5 cc/rev at 3000 psi

Main pump motor capacity: 7.5 kW/10 HP

Input supply to motor: 3-phase, 440 V, AC supply @ 50 Hz

Heat exchanger: Shell and tube heat exchanger

Heat transfer capacity: 45,360 kcal/h

Pressure medium: Servo 32

During operation of the system, the temperature of the oil increases. To maintain the oil temperature below a preset limit, one oil-to-water heat exchanger is integrated with the HPS. If the oil temperature exceeds the preset limit, a temperature sensitive switch mounted on the reservoir opens and turns off the HPS. A temperature gauge is fitted to the reservoir to monitor the oil temperature. The reservoir is equipped with a transparent oil level gauge to indicate the level and the relative contamination of hydraulic fluid in the reservoir. A low level switch is also mounted on the reservoir to turn off the HPS if the level of the fluid in the reservoir falls below the proper level. To maintain the quality of the oil coming out of the HPS within 3 μm, a filter of 3 μm is fitted at each outlet line prior to the reservoir.

3.3.2 Servo Valve with Manifold

The servo valves are the final control element in the closed loop servo hydraulic system and regulate the rate and direction of hydraulic fluid flow in the system. The servo valve is mounted on a manifold. It receives the control signal from the controller (see Figure 3.1). The control signal is a composite of the program command and the conditioned transducer feedback signal.

The servo valve converts this control signal into the physical movement of an internal spool, allowing the controlled porting of hydraulic fluid/oil to and from the actuator. The cross-sectional view of a servo valve is shown in Figure 3.2. The polarity of the control signal determines the direction of the spool and the amplitude of the control signal determines how far the spool will move, thus controlling the direction and rate of hydraulic fluid through the servo valve. As the amplitude of the control signal decreases to zero, the spool returns to its null position, thereby stopping the flow of the hydraulic fluid to and from the actuator.

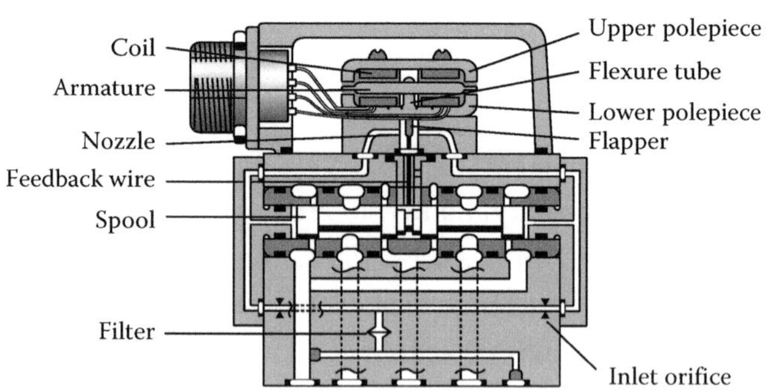

FIGURE 3.2
Cross-sectional view of servo valve. (Courtesy of Moog Inc., New York, NY, n.d., http://infokitabersama123.blogspot.in/2016/03/mengenal-prinsip-kerja-moog-servo-valve.html.)

3.3.3 Safety Relief Valve

Relief valves are provided for each channel in the system to set the maximum pressure limit in accordance with the desired pressure on the test component. This is provided for the safety of the test component. The pressure set value can be adjusted manually to the required value up to a maximum value of 200 bar.

3.3.4 Servo Controller

The servo controller provides the servo control, transducer conditioning, interlock, and readout functions for each channel of the closed loop electrohydraulic system. This controller is incorporated in such a way that the transducer conditioned output may be used only for readout, or it may be selected as feedback for the servo controller. The servo controller compares a composite command signal, which consists of the summation of an external program input signal to the controlled variable feedback signal. The composite command signal represents the desired amount and direction of a physical quantity to be applied by the hydraulic system. If the composite command and feedback are not equal, an error exists. By comparing these two signals, the servo controller develops an error signal that has a magnitude proportional to the difference between them and a polarity determined by their relative polarities. The error signal is applied to the valve amplifier which develops a valve drive signal to drive the servo valve that, in turn, controls the manifold opening. The polarity and magnitude of the valve drive signal cause the servo valve to open in the direction and amount required to allow or reduce the hydraulic oil, which reduces the error signal.

3.3.5 Pressure Transducer

The pressure transducer converts pressure into an analog electrical signal. Although there are various types of pressure transducers, the transducer which is selected for this pressure test facility is the strain-gauge transducer. The conversion of pressure into an electrical signal is achieved by the physical deformation of strain gauges which are bonded into the diaphragm of the transmitter and wired into a Wheatstone bridge configuration. Pressure applied to the pressure transducer produces a deflection of the diaphragm which introduces strain to the gauges. The strain will produce an electrical resistance change proportional to the pressure. Since a signal is required to be transmitted from the test component to the control room, which is located at an appreciable distance, a 4–20 mA output transducer made by Sensotec was selected for the application. A 4–20 mA signal is least affected by electrical noise and resistance in the signal wires over long-range signal transmission.

All wetted parts of the transducer are made of stainless steel and are hermetically welded. Therefore, there is no need for additional sealing material, which could possibly react with the pressure medium. The pressure transducer is periodically calibrated and its functionality within the accuracy level is ensured before the start of each test.

3.4 Integration of the Subsystems

All the subsystems of the facility are integrated by means of filling the test component with servo-32 oil and draining the oil back to the facility reservoir. The facility has been incorporated with a 1500 L reservoir for storing the hydraulic oil required for pressurization, as shown in Figure 3.3.

The specification of the test facility is as follows:

a. Maximum pressure: 200 bar
b. Accuracy: 0.5% of full scale
c. Number of channels: Two
d. Pressurizing medium: Servo system 32 hydraulic oil
e. Cycle profiles possible: Ramp, trapezoidal
f. Size of the test components: 700 mm diameter and 1500 mm long

The facility is incorporated with an oil level indicator and there is an oil temperature trip mechanism as an additional factor with regard to health and safety monitoring. The test facility is then audited by the representatives

FIGURE 3.3
The test facility.

from the Director General of Quality Assurance and is certified to conduct pressure tests to meet airworthiness requirements.

3.5 Design and Manufacturing of Adaptors

The cross-section and major dimensions of the test component are shown in Figure 3.4.

The assembly arrangement for conducting the pressure test is made as shown in Figure 3.5. Mild steel is selected as the material for all the adapting plates for the following reasons:

- Ease of availability
- Ease of machining
- Cost effectiveness
- Testing done at room temperature

However, engine standard fasteners are selected at the interface locations.

Rigid plates such as the bottom fixing plate, bottom sealing drum, inner bottom fixing plate, top sealing plate, and top holding plate are designed.

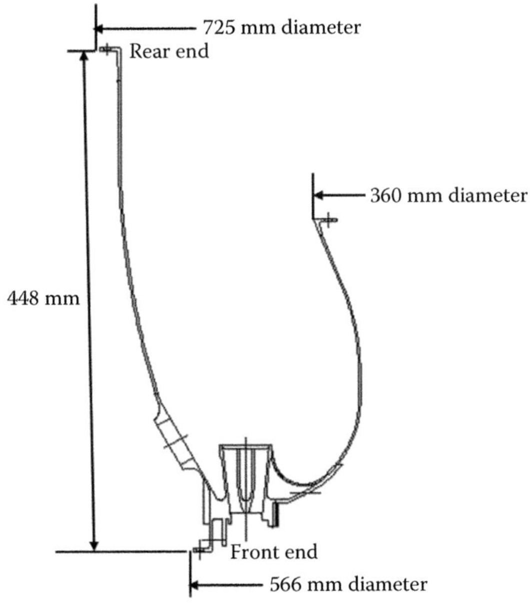

FIGURE 3.4
Cross-section of the combustor casing.

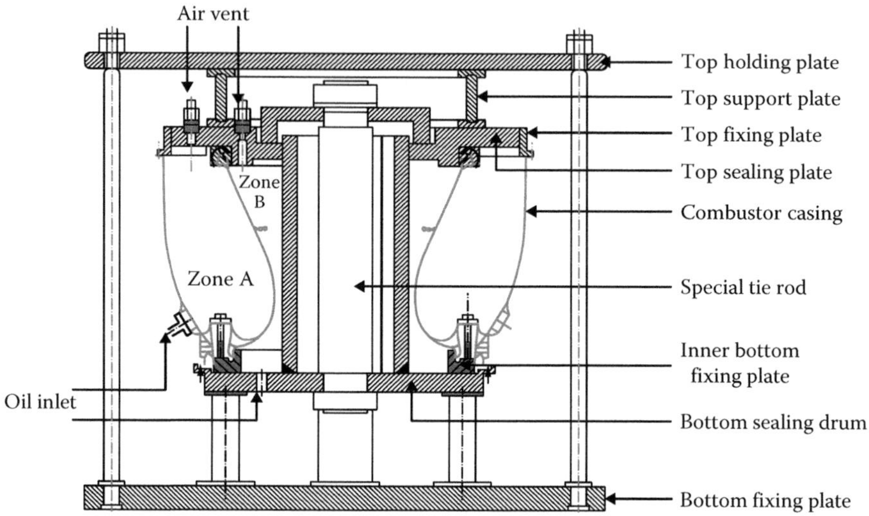

FIGURE 3.5
Assembly arrangement of the combustor casing. (From Sarangi, N. et al., *J. Test. Eval.*, 42(3), 629, 2014. With permission.)

The major dimensions of the adapting plates/structures are given in the following sections. The dimensions of the sealing plates are decided based on the dimensions of the combustor casing. Sealing plate thickness is kept high so that the adaptors should not fail during the test, causing failure of the test component.

3.5.1 Bottom Fixing Plate

This plate is designed for mounting the entire sealing assembly of the combustor casing. This is made of mild steel. The bottom fixing plate and its dimensions are shown in Figure 3.6. M24 tapped holes are provided to fix the tie rods, which are of 30 mm diameter. The tie rods are provided to take the axial load developed during pressurization.

3.5.2 Bottom Sealing Drum

The sealing drum, shown in Figure 3.7, is made out of mild steel to adapt the front end of the combustor casing.

The combustor front flange is 566 mm in diameter and has 80 holes of 6.5 mm diameter in a pitch circle diameter (PCD) of 554 mm. The same configuration is maintained at the interface location to fix the front flange of the combustor casing. The very basic purpose of this core is to guide and support the top sealing plate. Provision of this inner core also reduced the oil requirement by 25 L.

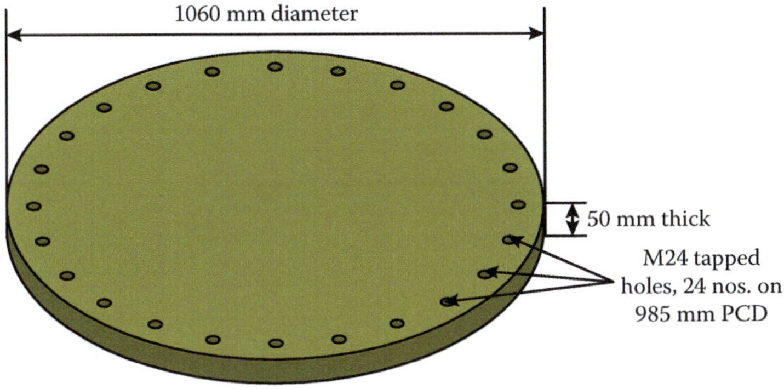

FIGURE 3.6
Bottom fixing plate.

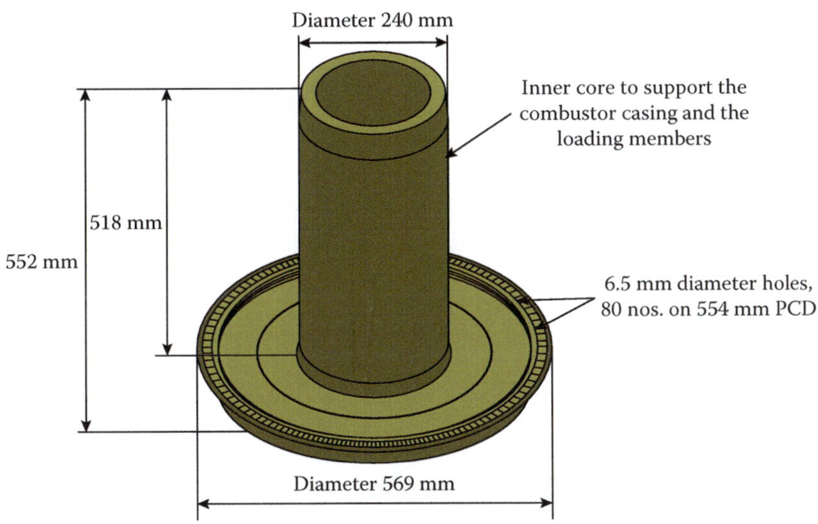

FIGURE 3.7
Bottom sealing drum.

3.5.3 Inner Bottom Fixing Plate

This plate was designed to provide sealing at the front end of the combustor casing (i.e., at the bottom of the sealing arrangement) and, at the same time, it prevents the interconnection between zone A and zone B. It is made out of mild steel, as shown Figure 3.8.

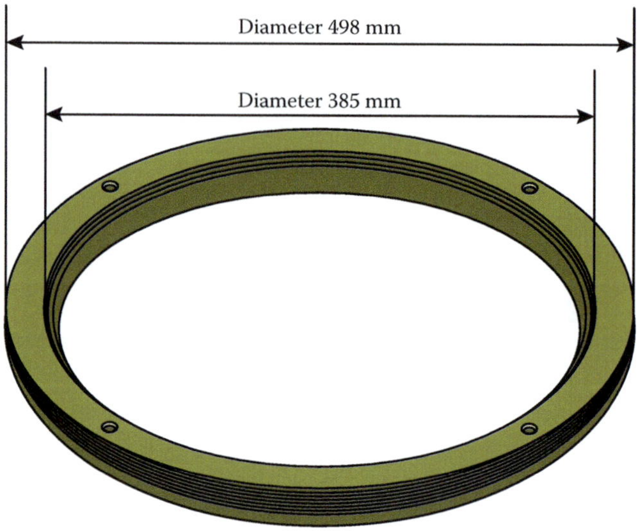

FIGURE 3.8
Inner bottom fixing plate.

3.5.4 Special Tie Rod

A special tie rod, 700 mm in length and 100 mm in diameter, was designed with M80 × 3 threads at both ends. The rod is located at the center of the assembly to constrain the top sealing plate in the axial direction. It is made out of mild steel and is shown in Figure 3.9.

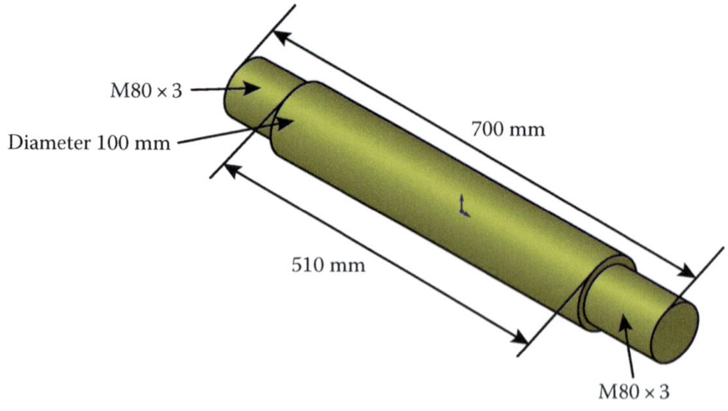

FIGURE 3.9
Special tie rod.

3.5.5 Top Sealing Plate

This plate is made out of mild steel and is designed to play the most important role during the pressure testing of a combustor casing. This plate has the provision of O-rings at three different diameter locations to provide better sealing. At the same time, it isolates zone B from zone A. This sealing arrangement enables the component to be tested as though in service. The component is able to float in the axial direction and there was no leakage of the oil due to its having been sealed by O-rings. The plate is supported on a step provided on the bottom sealing drum. Its upward movement is arrested by the top support plate. The top sealing plate is shown in Figure 3.10.

3.5.6 Top Fixing Plate

This plate is designed to suit the rear end of the combustor casing. This plate has a flange similar to the combustor casing with 106 holes of 7.6 mm diameter in a PCD of 703 mm. This plate is made annular to the top sealing plate in such a way that the top fixing plate, along with the rear flange, is free in the axial direction. It is made out of mild steel and is shown in Figure 3.11.

3.5.7 Top Support Plate

The top support plate is designed to play the role of a spacer ring between the top sealing plate and the top holding plate. It is made of mild steel and has an I section for rigidity. The load is transferred from the top sealing plate

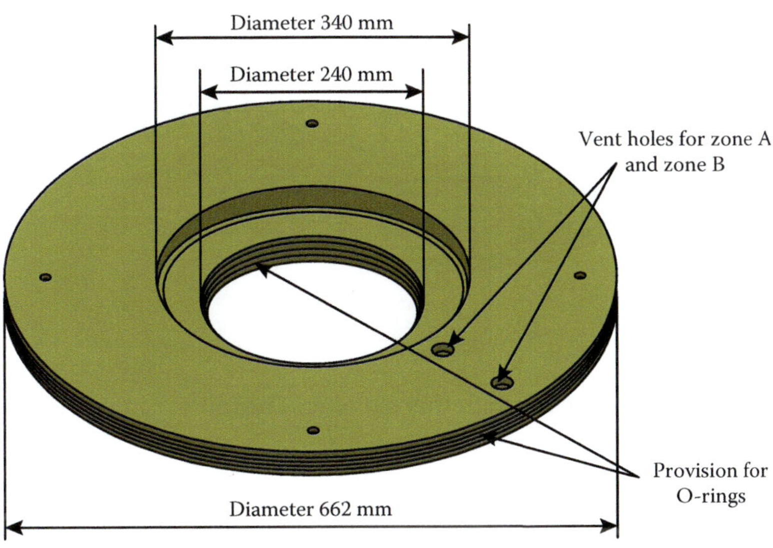

FIGURE 3.10
Top sealing plate.

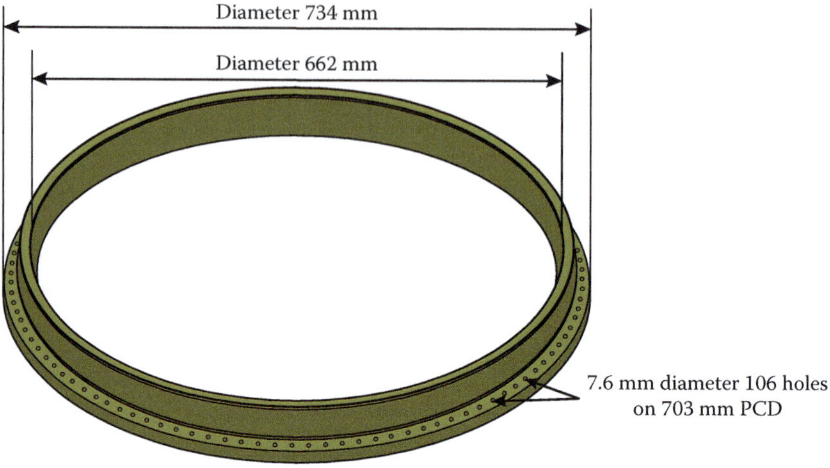

FIGURE 3.11
Top fixing plate.

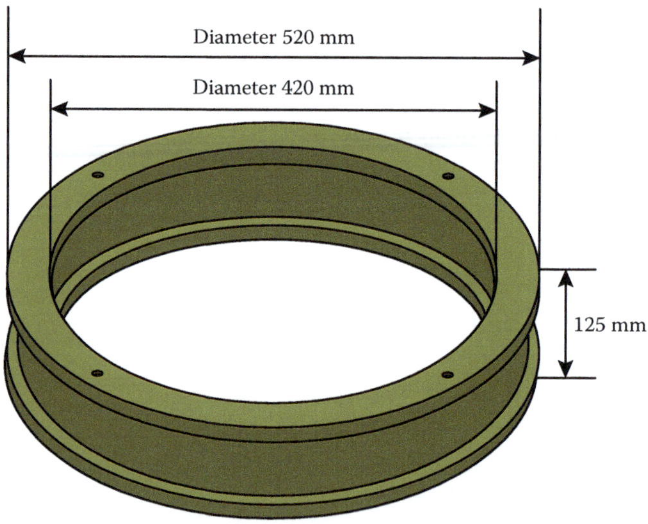

FIGURE 3.12
Top support plate.

to the top holding plate through this member. The top sealing plate is shown in Figure 3.12.

3.5.8 Top Holding Plate

This plate is designed to prevent the axial movement of the top sealing plate as well as the top support plate. It is made of 25 mm thick mild steel plate and

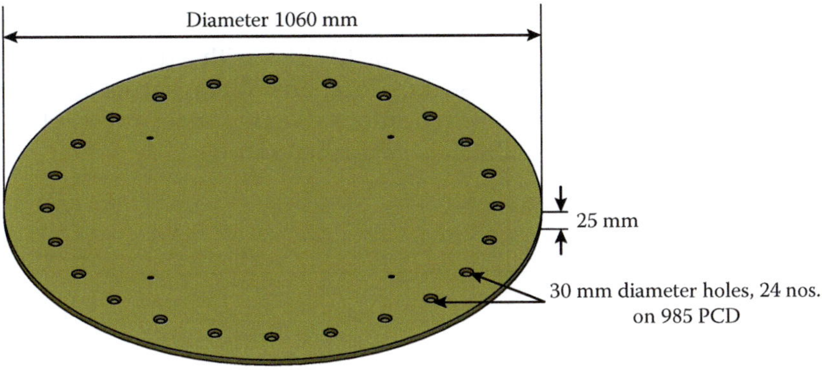

FIGURE 3.13
Top holding plate.

is 1060 mm in diameter, and 24 holes of 30 mm diameter have been provided in a PCD of 985 mm so that it can be held against the bottom fixing plate by means of 24 tie rods of 30 mm diameter. The top holding plate is shown in Figure 3.13.

3.5.9 Special Studs

During pressurization, the top sealing plate is subjected to a load of about 100 tons, which need to be arrested. This is done partly by the special stud at the center and partly by the special studs at the circumference; 12 special studs are used for this purpose, though provision is made for 24. The studs are 1065 mm in length with a 30 mm diameter and are machined out of mild steel. The special stud is shown in Figure 3.14.

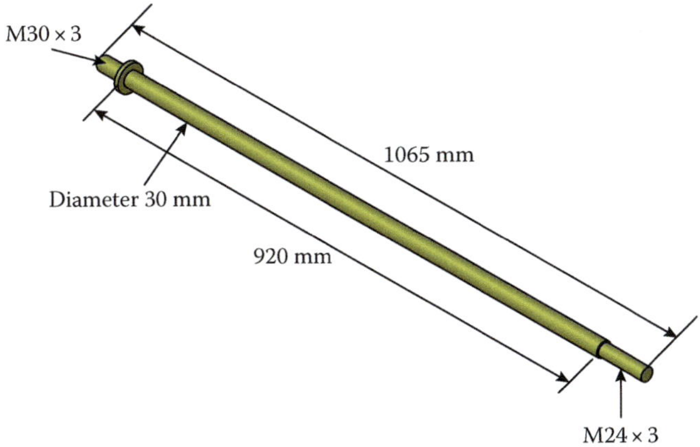

FIGURE 3.14
Special stud.

3.5.10 Seals

All the interface locations are sealed using O-rings with a diameter of either 2.5 or 3.5 mm depending on the sealing locations. Teflon sheets are used to seal the flat surfaces at the ignitor locations. Bonded seals are used at the blanking locations to prevent oil leakage during testing.

3.6 Summary

A unique and complex state-of-the-art test facility and test setup were developed indigenously to conduct pressure tests on the combustor casing. All the adapting components were designed, manufactured, and inspected with respect to the drawing dimensions, examined for cracks on the surface, and were accepted. Engine standard fasteners, required at the interface locations, were used during the test. The test facility and the test setup were audited by a third-party organization and were certified prior to start of the test.

The combustor casing was instrumented and was assembled with the adaptors in the test setup. The test component was filled with hydraulic oil through the inlet ports. Zones A and B were individually pressurized to ensure the component was leak proof between the zones, as well as at the boss locations. Several test cycles were carried out at a low pressure of about 15 bar and the test setup was maintained ready to conduct the cyclic pressure test on the combustor casing in accordance with the program requirements. Details of the test are described in Chapter 4.

4

Manufacturing of an Aero Engine Combustor Casing, the Experimental Evaluation of Its Fatigue Life, and Correlation with Numerical Results

4.1 Introduction

The structural design of annular combustors for aero gas turbines offers substantial challenges to engine development laboratories due to the functional aspects, such as complex geometry, thin wall structures, and stringent airworthiness requirements. As a combustor is a critical component of a gas turbine engine, its structural integrity is paramount to ensure the safety of engine operation. Compliance with stringent military airworthiness standards is essential for the certification of combustors for use in aero gas turbines and this chapter describes the method of integrating the physical tests and finite element simulation in order to ensure safe operation of the combustor throughout its intended life. The focus of the current study is the evaluation of the fatigue life of an annular combustor used in an afterburning turbofan engine producing 80 kN thrust. The intended use of this engine is the propulsion of a lightweight multirole combat aircraft with the ability to achieve supersonic speeds in the order of Mach 2. During engine operation, the combustor casing experiences pressure differentials of more than 25.3 bar and military standards stipulate endurance of the component on being subjected to 7000 pressurization cycles with peak pressure levels of 30 bar. This chapter describes the testing of the annular combustor in a custom-developed closed loop hydraulic test facility and reports the results corresponding to cyclic testing. The experimental strain levels that are monitored through various strain gauge installations are compared with those of numerical procedures. The correlated data is subsequently used for the redesign of several combustor features leading to critical stress mitigation without compromising component safety. This chapter presents a novel approach wherein the experimental results, in tandem with numerical estimates, are used not only for ensuring the structural integrity of the component in accordance with military airworthiness standards, but also for design iterations leading to fatigue life enhancement.

4.2 Manufacturing Method of the Combustor Casing

The combustor casing under investigation in the present study is manufactured from a circular metal blank of 2 mm thickness using the metal spinning process. The flanges are machined out of forgings of the same material and then welded circumferentially with the spun component. The centrifugal spinning process is described in detail in the following section.

4.2.1 Metal Spinning Process

Metal spinning is the process of forming three-dimensional symmetrical parts from flat circles of metal. Metal spinning provides an economical alternative to stamping, casting, and many other metal-forming processes. It is an excellent means of quickly prototyping round hollow metal forms (primarily the realm of expensive sheet metal stamping machinery). Metal spinning allows the production of hollow, axially symmetrical sheet metal components. The basic technique of spinning, which is common to this process, consists of clamping a sheet metal blank against a mandrel on a spinning lathe, and gradually forming the blank onto the mandrel surface with a roller, either in a single step or series of steps [113–115]. The profile of the mandrel is the ID profile of the desired finished part. A levered force is applied uniformly to the sheet metal by rotating the metal and its intended form (the mandrel) at very high rotational speeds, thus the sheet metal is deformed evenly without any wrinkling. The spinning process allows for the rapid production of multiple parts as well as quick reiteration since only the one tool (the mandrel) need be modified. The metal spinning process is explained in Figures 4.1 and 4.2.

The mandrel is typically referred to as a "spin chuck" and is usually made from wood or steel. Wood tooling is often used for short-run or prototype work. Steel tooling is used when tighter tolerances are required and for longer running production runs. Depending on the complexity of the part being spun, spinning can be highly demanding physically [116,117]. The interior surface (against the mandrel) becomes as smooth as the surface of the mandrel. A lubricating wax or grease is essential to achieving a quality finish and being able to easily remove the part from the mandrel. Metals harden as they are worked that sometimes necessitates annealing the piece partway through a spin, but this is not often necessary and the metal hardens to a desirable stiffness as the part is spun.

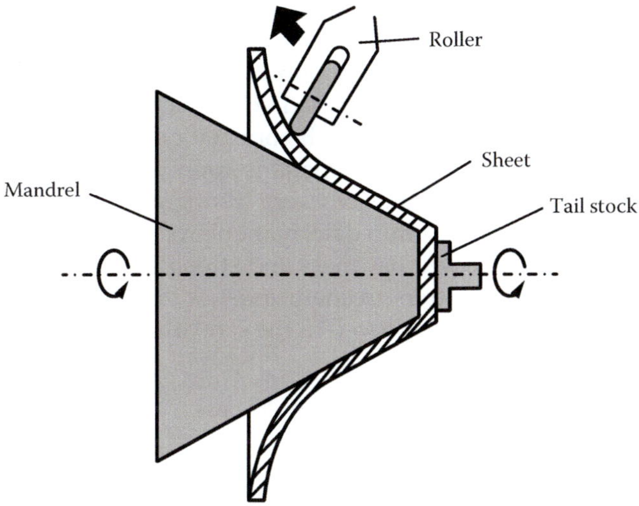

FIGURE 4.1
The metal spinning process. (Courtesy of Low Carbon and Materials Processing Group, Cambridge University, Department of Engineering, Cambridge, U.K., n.d., http://www.lcmp.eng.cam.ac.uk/welcome/equipment/flexible-asymmetric-spinning.)

FIGURE 4.2
The sequence of operations in a metal spinning process: (a) blank ready to spin, (b) spinning, and (c) finished part.

4.2.1.1 Advantages of the Metal Spinning Process

With metal spinning, lead times are usually shorter and tooling costs are significantly lower than with other metal-forming processes such as stamping. Lower volume production runs can be economically produced with centrifugal spinning. Almost every metal that is available in sheet form may be spun. Reverse angles can be achieved with metal spinning. The many design possibilities and cost advantages make centrifugal spinning an attractive process to consider for manufacturing precision products.

Metals are crystalline in nature and consist of irregularly shaped grains of various sizes. Each grain is made up of atoms in an orderly arrangement, known as a "lattice." The orientation of the atoms in a grain is uniform but differs in adjacent grains. When a force is applied to deform it or change its shape, many changes occur in the grain structure. These include grain fragmentation, movement of atoms, and lattice distortion. Slip planes develop through the lattice structure at points where the atom bonds of attraction are the weakest and whole blocks of atoms are displaced. The orientation of atoms, however, does not change when slip occurs. When metal is spun in a cold state, there is no recrystallization of grains and thus recovery from grain distortion or fragmentation does not take place. As grain deformation proceeds, greater resistance to this action results in increased hardness and strength. The metal is said to be strain hardened. Thus, the strength characteristics in the direction of spin and the direction of forming are improved in the centrifugal spinning process.

4.2.1.2 Disadvantages of Metal Spinning Process

Though the components manufactured in this way incur a lower tooling cost and are economical, it is extremely difficult to maintain the thickness of the manufactured part throughout within the design limits. This variation of thickness leads to the mismatch of the bosses with the parent sheet metal, which reduces the effective thickness for load transfer.

If a crack forms or the object is dented during the spinning process, it must be scrapped. Repairing the object is not cost-effective. This scrappage adds to the production cost, but is less onerous than the problems that may be encountered during operation should a part prove faulty.

Being an experimental engine, there are several bosses welded to its surface, sometimes in the order of as many as 54. The component has circumferential welds at three locations. The circumferential welding is carried out by electron beam welding (EBW) whereas the bosses are welded to the sheet metal body by tungsten inert gas (TIG) welding. The details of the welding as explained in the next section.

4.2.2 Electron Beam Welding Method

EBW is a fusion welding process in which a beam of high-velocity electrons is applied to two materials to be joined. The workpieces melt and flow together as the kinetic energy of the electrons is transformed into heat upon impact. EBW is often performed in a vacuum in order to prevent dissipation of the electron beam. It was developed by German physicist Karl-Heinz Steigerwald who, at the time, was working on various electron beam applications; Steigerwald conceived and developed the first practical EBW machine, which began operation in 1958 [119–122].

Not all materials may be welded by electron beam in a vacuum. This technology cannot be applied to materials with high vapor pressure at the melting temperature, such as zinc, cadmium, magnesium, and practically all nonmetals. Another limitation of weldability may be the change of material properties inflicted by the welding process; for example, the high speed of cooling.

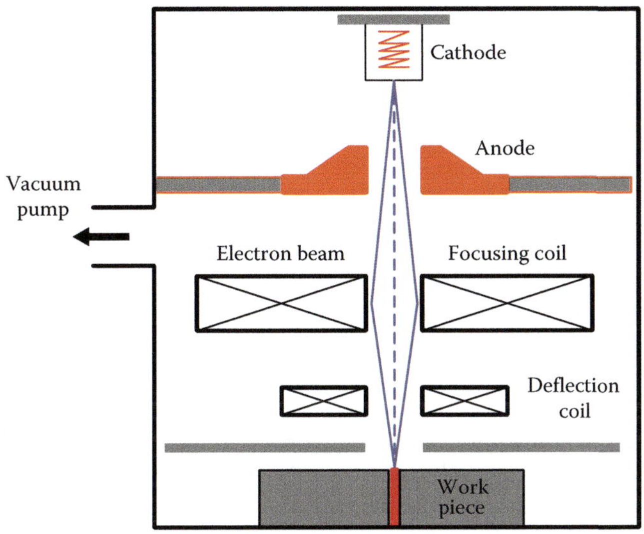

FIGURE 4.3
Schematic of electron beam welding equipment. (Courtesy of Dmitri Kopeliovich, Electron Beam Welding (EBW), n.d., http://www.substech.com/dokuwiki/doku.php?id=electron_beam_welding_ebw.)

Electron beam equipment comprises the following (see Figure 4.3):

- Electron gun, generating the electron beam (cathode)
- Working chamber, mostly evacuated to "low" or "high" vacuum
- Workpiece manipulator (positioning mechanism)
- Deflection coil
- Supply and control/monitoring electronics

Advantages of EBW:

- It produces a weld of superb quality, with extremely deep penetration, while at the same time minimizing overall heat input.
- EBW can even weld extremely thin materials with ease.
- It can also weld thicker sections in a single pass.
- It can be possible to achieve a nearly 0% reject rate (with strict management and knowledgeable operators).

Disadvantages of EBW:

- EBW is by far the most costly welding process; the equipment can cost hundreds of thousands to millions of dollars.
- EBW requires a vacuum chamber containing a hard vacuum.

- Only small to medium-sized items can be welded.
- Though the welding itself can be done very swiftly, overall EBW is time-consuming.
- The equipment is complex and there are quite a few process variables involved.

4.2.3 Tungsten Inert Gas Welding Method

TIG welding or gas tungsten arc welding are arc welding processes that use a nonconsumable tungsten electrode to produce the weld. The weld area is protected from atmospheric contamination by an inert shielding gas (argon or helium) and a filler metal is normally used, though some welds do not require it. A constant-current welding power supply produces energy which is conducted across the arc through a column of highly ionized gas and metal vapors known as a "plasma" [124–127]. TIG welding is most commonly used to weld thin sections of stainless steel and nonferrous metals such as aluminum, magnesium, and nickel and copper alloys. The process grants the operator greater control over the weld than competing processes such as shielded metal arc welding and gas metal arc welding, allowing for stronger, higher quality welds. However, TIG is comparatively more complex and difficult to master and, furthermore, it is significantly slower than most other welding techniques. A related process, plasma arc welding, uses a slightly different welding torch to create a more focused welding arc and, as a result, is often automated. A TIG welding process is shown in Figure 4.4.

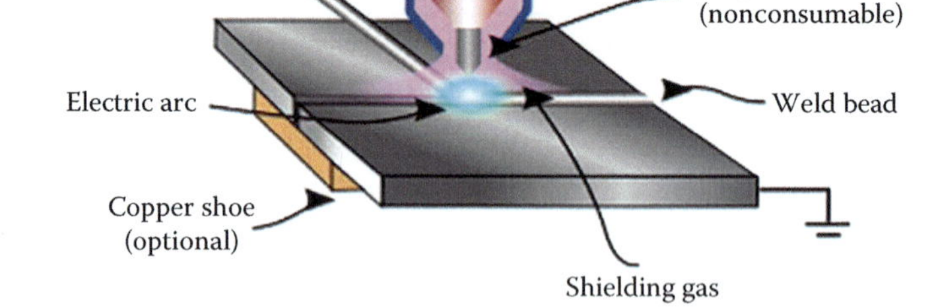

FIGURE 4.4
Tungsten inert gas welding process. (Courtesy of Euroweld: Used Welding and Manipulation Equipment, Euroweld: Your partner in professional used welding equipment, GS Oud Gastel, the Netherlands, n.d., http://usedwelding-equipment.com/tig-welding/.)

- *Advantages of TIG welding*
  - TIG welding produces high-quality welds.
  - The weld is automatically protected by the inert gas during the welding process.
  - No slag is produced.
  - TIG welding can be done in any position.
- *Disadvantages of TIG welding*
  - TIG welding is a slow process.
  - Highly skilled labor is needed.
  - Welder is exposed to huge intensities of light.

Stresses induced due to spinning are relieved by annealing and the formed part is welded to the flanges at both ends by the EBW process. Parameters such as pressure and temperature are monitored at several axial and circumferential locations of the combustor, which necessitates the mounting of nearly 50 instrumentation bosses to the combustor surface to mount atomizers, igniters, light up atomizers, boroscopes, and other instrumentation probes. The bosses are welded to the wall of the combustor casing, leading to the development of residual stresses. The entire assembly is subjected to post-weld heat treatment (PWHT) to reduce and redistribute the residual stresses before releasing it for use in the engine assembly. The PWHT restores the macro structure and also prevents the occurrence of brittle fractures on the component during its application.

4.3 Configuration of the Combustor Casing

A combustor casing consists of inner and outer segments, flanges, and numerous bosses for the mounting of pressure probes, fuel atomizers, and an igniter, as shown in Figures 4.5 and 4.6. It has a high-pressure main chamber (zone A) wherein the combustion takes place and an inner chamber (zone B) through which the transmission shaft passes, with the pressure differential between the two reaching about 10 bar during engine operation [129].

The combustor casing serves as a pressure vessel that must withstand this pressure differential with a mechanical load, rather than a thermal load, as the driving design factor. Space and weight are at a premium in aircraft applications and, hence, a well-designed combustor strives to be compact and lightweight. These operational requirements entail the manufacturing of combustor casings out of thin sheets made of high temperature resistant nickel super alloy.

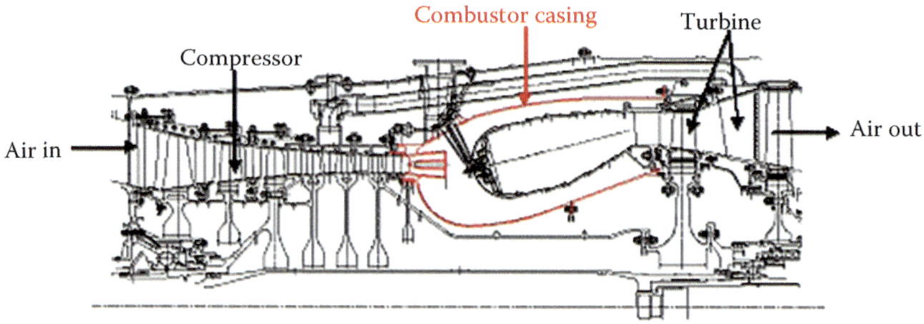

FIGURE 4.5
Schematic of the combustor casing in the engine. (From Panigrahi, S.K., Sarangi, N., and Chandrasekhar, U.: Experimental evaluation of overload capability of an annular combustor casing of a gas turbine engine. *J. Exp. Tech.* 2015. 40. 841–848. Copyright Wiley-VCH Verlag GmbH & Co. KGaA. Reproduced with permission.)

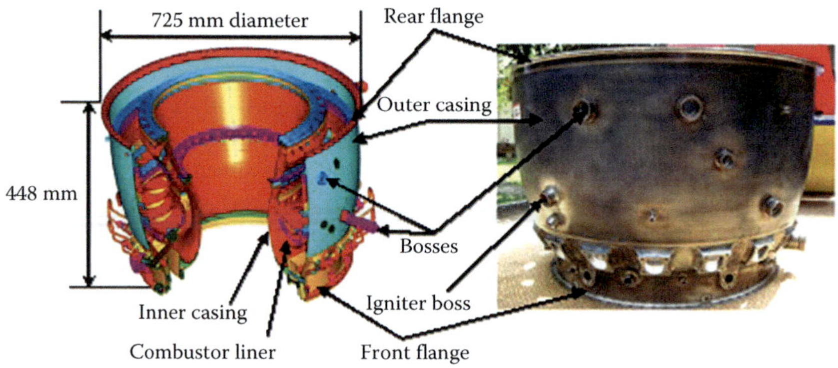

FIGURE 4.6
Configuration of the combustor casing. (From Panigrahi, S.K., Sarangi, N., and Chandrasekhar, U.: Experimental evaluation of overload capability of an annular combustor casing of a gas turbine engine. *J. Exp. Tech.* 2015. 40. 841–848. Copyright Wiley-VCH Verlag GmbH & Co. KGaA. Reproduced with permission.)

4.4 Experimental Evaluation of Fatigue Life

The combustor casing under consideration in the present study is intended to withstand 7000 zero–maximum–zero pressure cycles where the maximum pressure is equal to the maximum operating pressure in the engine. The primary objective of the experimental testing of the combustor casing is to evaluate its structural integrity when it is subjected to 7000 cycles of pressurization and pressure release at operational temperatures varying from 350°C to 600°C. The maximum pressure experienced by the combustor casing is 25.3 bar in the engine. The inner casing experiences 16.6 bar. An indigenously developed servo hydraulic test facility working on a closed loop with two independent pressurization channels is used to conduct the cyclic pressure test (see Chapter 3).

The testing is proposed to be conducted at room temperature instead of at a higher temperature and, to account for temperature variation, the pressure differential is increased to 29.8 bar in the main chamber and 19.0 bar in the inner chamber by multiplying with a compensation factor of 1.17. This compensation factor is determined through an empirical relation that accounts for the loss of material strength at high temperatures. This is the typical practice of temperature compensation used by leading aero engine designers [130,131].

4.4.1 Instrumentation

The accurate measurement of strain, from which the stress can be determined, is one of the most significant predictors of fatigue life. Prediction of fatigue life often requires the experimental measurement of localized loads, the frequency of the load occurrence, the statistical variability of the load, and the number of cycles a part will experience at any given load. A variety of methods may be used to predict the fatigue life by applying either a linear or a weighted response to the measured parameters. Experimental measurements are made to determine the minimum and maximum values of the load over a time period adequate to establish the repetition rate. If the part is complex in shape, such that the strain levels cannot be easily or accurately predicted from the loads, strain gauges will need to be applied to the component in critical areas. Fatigue life prediction is based on knowledge of both the number of cycles the part will experience at any given stress level during that life cycle and other influential environmental and use factors. One of the most commonly accepted methods of measuring strain is the resistive strain gauge. Modern strain gauges are resistive devices that experimentally evaluate the load or the strain an object experiences.

In the present study, electrical resistance strain gauges are bonded to the combustor casing to monitor for strain signals. The strain gauges are mounted at predefined locations, as shown in Figure 4.7, to enable close monitoring of the strains during the test required to ensure the health and safety of the test facility. Pressure transducers, one in each zone, are mounted to monitor the pressure signals as feedback to the control system [132–134]. The signals from the pressure transducers are fed to the controller and then to the computer.

The pressure cycle used for the fatigue testing of the combustor casing is shown in Figure 4.8. The cycle is defined in the control system computer which is connected to the servo controller of the test facility. The combustor

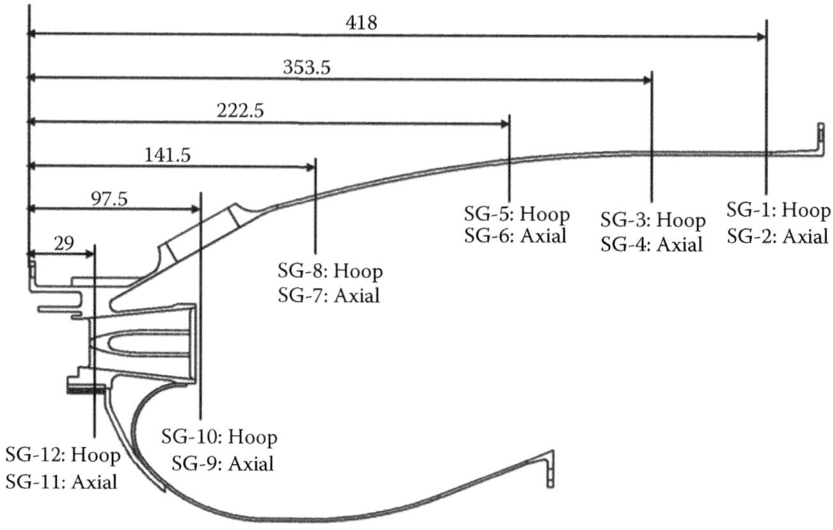

FIGURE 4.7
Location of strain gauges on the combustor casing (dimensions are in mm). (From Panigrahi, S.K., Sarangi, N., and Chandrasekhar, U.: Experimental evaluation of overload capability of an annular combustor casing of a gas turbine engine. *J. Exp. Tech.* 2015. 40. 841–848. Copyright Wiley-VCH Verlag GmbH & Co. KGaA. Reproduced with permission.)

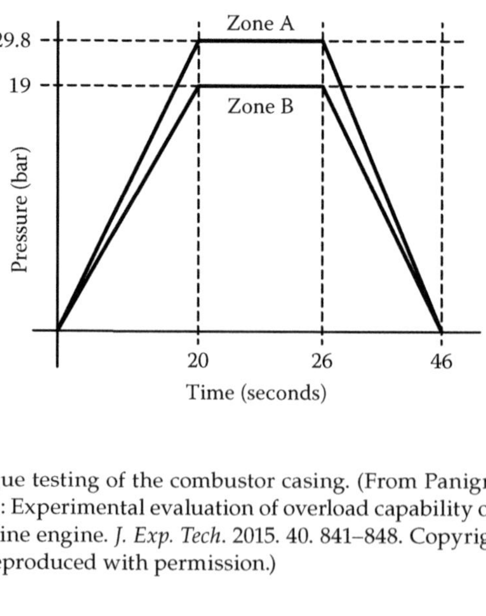

FIGURE 4.8
Pressure cycle for fatigue testing of the combustor casing. (From Panigrahi, S.K., Sarangi, N., and Chandrasekhar, U.: Experimental evaluation of overload capability of an annular combustor casing of a gas turbine engine. *J. Exp. Tech.* 2015. 40. 841–848. Copyright Wiley-VCH Verlag GmbH & Co. KGaA. Reproduced with permission.)

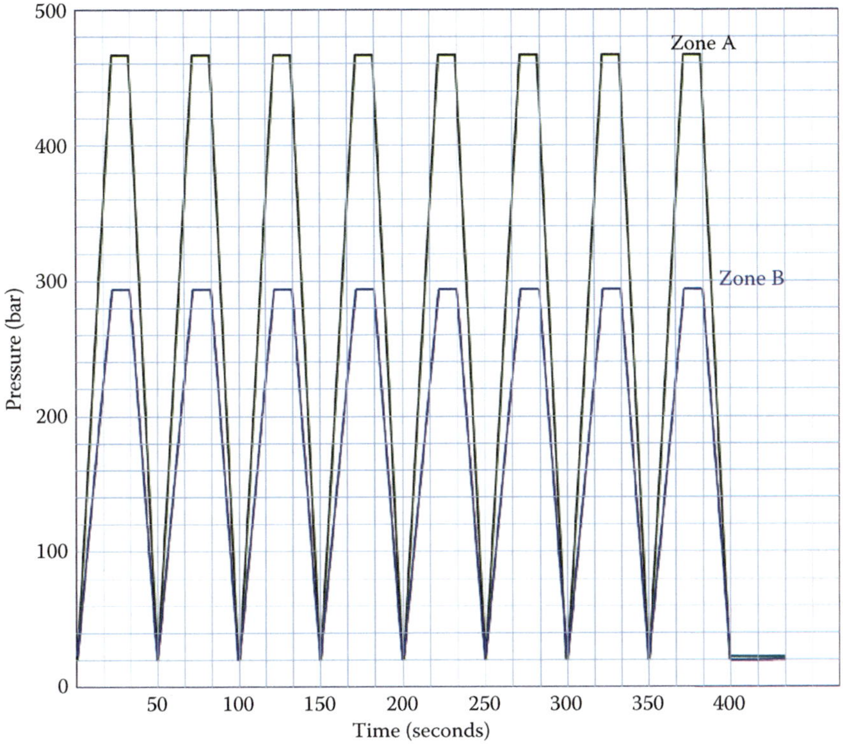

FIGURE 4.9
The pressure plot taken during the test. (From Panigrahi, S.K., Sarangi, N., and Chandrasekhar, U.: Experimental evaluation of overload capability of an annular combustor casing of a gas turbine engine. *J. Exp. Tech.* 2015. 40. 841–848. Copyright Wiley-VCH Verlag GmbH & Co. KGaA. Reproduced with permission.)

casing is subjected to this pressure cycle. The pressure values experienced by the transducers are superimposed on the defined pressure cycle to ensure that the component is exposed to the defined pressure. The test procedure involves inspection on completion of every block of 1000 pressurization cycles.

The pressure plot obtained from the test rig is shown in Figure 4.9. A cyclic pressure test is conducted on the combustor casing at the end of 1000 cycles.

4.4.2 Inspection Methodology

It is regular practice that a nondestructive technique is followed to carry out inspection of critical aero engine components at various stages of testing.

Nondestructive testing is one part of the function of quality control and is complementary to other long-established methods. By definition, nondestructive testing is the testing of materials, for surface or internal flaws or for their metallurgical state, without interfering in any way with the integrity of the material or its suitability for service.

The technique can be applied on a sampling basis for individual investigation or may be used for 100% checking of material in a production quality control system. While being a high technology concept, evolution of the equipment has made it robust enough for application in any industrial environment at any stage of manufacture—from steel-making to site inspection of components already in service. A certain degree of skill is required to apply the techniques properly in order to obtain the maximum amount of information concerning the product, with consequent feed back to the production facility.

Nondestructive testing is not just a method for rejecting substandard material; it is also an assurance that the supposedly good is good. The technique uses a variety of principles; there is no single method around which a black box may be built to satisfy all requirements in all circumstances.

Out of the varieties of method available, the following nondestructive inspection methods are carried out at the end of 1000 pressure cycles:

- Fluorescent particle inspection to reveal the presence of any cracks on the surface
- Radiography at the weld locations to reveal the presence of any subsurface cracks
- Dimensional inspection at several predefined locations

The component was found to be free of defects and testing was continued further. After completion of 1825 cycles, oil was found to be leaking from an ignitor boss-shell weld location in the component, as shown in Figure 4.10, indicating a premature failure and the test was stopped immediately. The test was aborted and dye penetrant inspection was carried out in situ, which confirmed the crack on the weld surface. A detailed failure investigation was carried out and a crack of 30 mm in length was found along the weld surrounding the boss. Based on the failure analysis, it was inferred that the primary cause for premature structural failure was stress concentration near the ignitor boss; this entailed a design modification.

An attempt was made to carry out structural analysis of the combustor casing followed by numerical analysis of the critical area, and to make a correlation between the experimental and analytical results. The mechanical properties of the Inconel 718 sheet are evaluated by the tensile testing of flat specimens, as discussed in the following section.

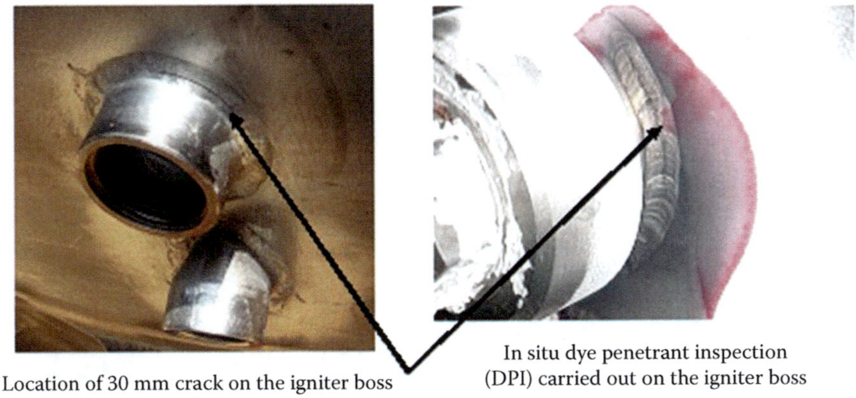

FIGURE 4.10
The igniter boss where the crack is observed during testing. (From Panigrahi, S.K., Sarangi, N., and Chandrasekhar, U.: Experimental evaluation of overload capability of an annular combustor casing of a gas turbine engine. *J. Exp. Tech.* 2015. 40. 841–848. Copyright Wiley-VCH Verlag GmbH & Co. KGaA. Reproduced with permission.)

4.5 Mechanical Properties of Combustor Casing Material

To evaluate the material properties of the Inconel 718 in sheet configuration, five test specimens were prepared in accordance with American Society for Testing and Materials (ASTM) standards, as shown in Figure 4.11, from the same batch of Inconel sheets and were tested in the universal testing machine, as shown in Figure 4.12 [135,136]. The ultimate tensile strength (UTS) was found to be 1281 MPa and the percentage reduction in the area was found to be 15%.

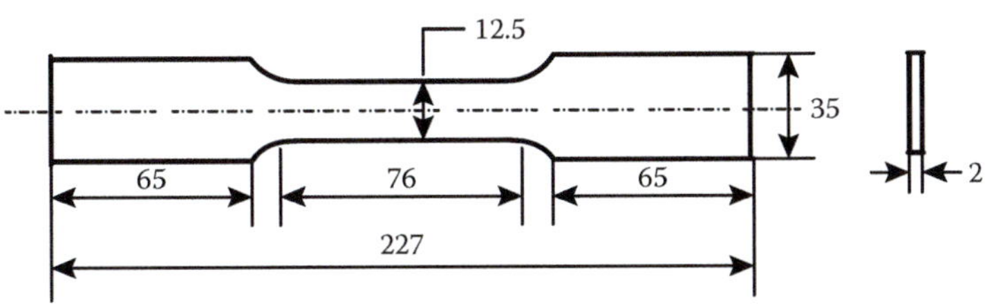

FIGURE 4.11
Test specimen (dimensions are in mm). (From Panigrahi, S.K., Sarangi, N., and Chandrasekhar, U.: Experimental evaluation of overload capability of an annular combustor casing of a gas turbine engine. *J. Exp. Tech.* 2015. 40. 841–848. Copyright Wiley-VCH Verlag GmbH & Co. KGaA. Reproduced with permission.)

FIGURE 4.12
The universal testing machine. (From Panigrahi, S.K., Sarangi, N., and Chandrasekhar, U.: Experimental evaluation of overload capability of an annular combustor casing of a gas turbine engine. *J. Exp. Tech.* 2015. 40. 841–848. Copyright Wiley-VCH Verlag GmbH & Co. KGaA. Reproduced with permission.)

4.6 Numerical Analysis of the Combustor Casing

Numerical analysis was carried out in accordance with the engine condition of the combustor casing using the finite element method, as shown in Figure 4.13.

The objective of this analysis is to estimate the stresses in the combustor casing in the maximum operating condition and to evaluate the hoop stress at the failed area. Aerodynamic pressure and thermal loads corresponding to the maximum engine operating conditions were considered for analysis.

Inconel 718 [25] is the material used for manufacturing the combustor casing. The temperature-dependent mechanical properties of Inconel 718 are given in Tables 4.1 and 4.2.

The finite element model of the casing was generated using Hypermesh software. Four-noded shell elements were used for the skin of the combustor, whereas eight-noded hexahedron elements were used for thick sections, such as the atomizer boss locations. Various adaptors and bosses on the combustor casing are modeled in a detailed manner. The finite element

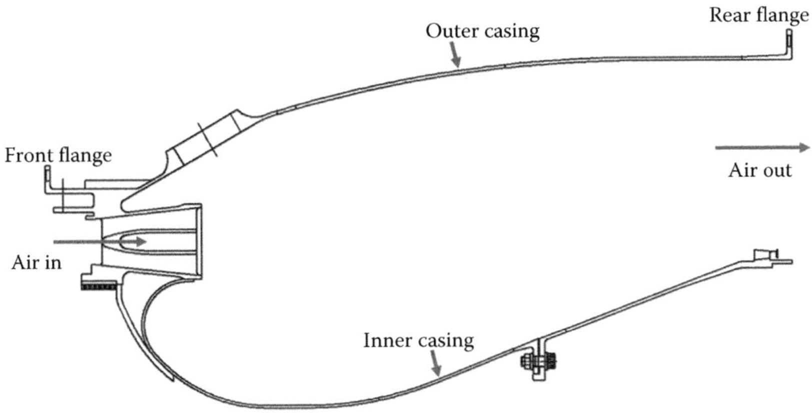

FIGURE 4.13
Schematic of combustor casing considered for numerical analysis. (From Panigrahi, S.K., Sarangi, N., and Chandrasekhar, U.: Experimental evaluation of overload capability of an annular combustor casing of a gas turbine engine. *J. Exp. Tech.* 2015. 40. 841–848. Copyright Wiley-VCH Verlag GmbH & Co. KGaA. Reproduced with permission.)

TABLE 4.1
Mechanical Properties of Inconel 718

Temperature (°C)	Modulus of Elasticity (MPa)	Coefficient of Thermal Expansion, α (mm/mm °C (10^{-6}))	Poisson's Ratio (μ)
20	200,098.49	12.150	0.294
100	195,537.82	12.218	0.287
200	190,572.00	13.238	0.280
300	184,916.53	13.702	0.273
400	179,261.07	14.044	0.272
500	166,518.86	14.308	0.272
600	166,488.45	14.672	0.277

TABLE 4.2
Thermal Conductivity of Inconel 718

Temperature (°C)	21	93	204	316	427	538	649	760
$K \times 10^{-3}$ (W/mm °C)	11.4	12.6	14.14	16.02	17.9	19.6	21.4	23.2

model is shown in Figure 4.14. The coordinates for the model are shown in Figure 4.15.

4.6.1 Details of the Finite Element Model

Number of nodes in the model: 136,810

Number of elements in the model: 123,432

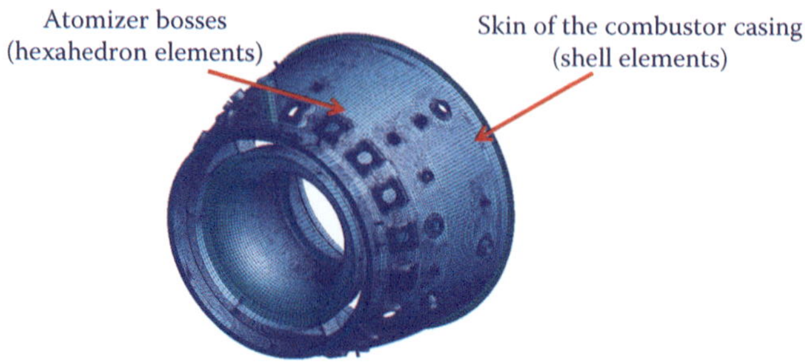

FIGURE 4.14
Finite element model of the combustor casing. (From Panigrahi, S.K., Sarangi, N., and Chandrasekhar, U.: Experimental evaluation of overload capability of an annular combustor casing of a gas turbine engine. *J. Exp. Tech.* 2015. 40. 841–848. Copyright Wiley-VCH Verlag GmbH & Co. KGaA. Reproduced with permission.)

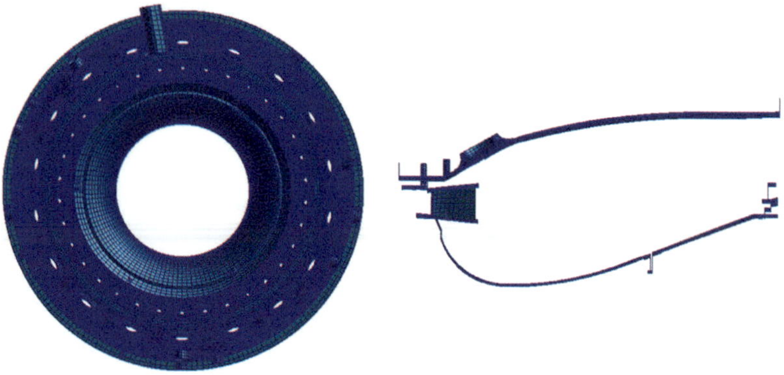

FIGURE 4.15
Global cylindrical co-ordinate system. (From Panigrahi, S.K., Sarangi, N., and Chandrasekhar, U.: Experimental evaluation of overload capability of an annular combustor casing of a gas turbine engine. *J. Exp. Tech.* 2015. 40. 841–848. Copyright Wiley-VCH Verlag GmbH & Co. KGaA. Reproduced with permission.)

4.6.2 Finite Element Model Quality Parameters

Aspect ratio = 4 (>98% of the elements meet the requirement and the elements that do not meet the requirement are located away from the high-stress zone).

Warp = 6 (>90% elements meet the requirement and the elements that do not meet the requirement are located away from the high-stress zone).

Jacobian = 0.7 (>97% elements meet the requirement and the elements that do not meet the requirement are located away from the high-stress zone).

4.6.3 Boundary Conditions

The combustor was bolted circumferentially at the front flange of the combustor casing to the rear flange of the compressor casing. Single point constraints were used for finite element simulation by constraining the axial and tangential degrees of freedom of nodes at the front flange. The front and rear inner casings were bolted to each other with multipoint constraints (MPC), as shown in Figure 4.16.

4.6.4 Load

Analysis was carried out considering the pressure and thermal loads on the combustor casing for the maximum operating conditions.

4.6.5 Pressure Load

The maximum differential pressure applied to the outer casing was 25.3 bar and the maximum differential pressure applied to the inner casing was 8.7 bar.

4.6.6 Thermal Load

The outer casing of the combustor is cooled by low-pressure compressor exit bypass air, as shown in Figure 4.17. The inner casing is cooled by the

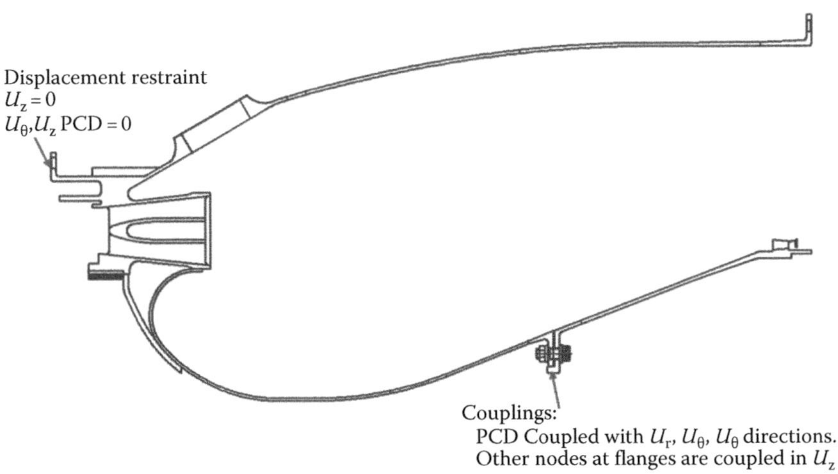

FIGURE 4.16
Constraints applied on the FE model. (From Panigrahi, S.K., Sarangi, N., and Chandrasekhar, U.: Experimental evaluation of overload capability of an annular combustor casing of a gas turbine engine. *J. Exp. Tech.* 2015. 40. 841–848. Copyright Wiley-VCH Verlag GmbH & Co. KGaA. Reproduced with permission.)

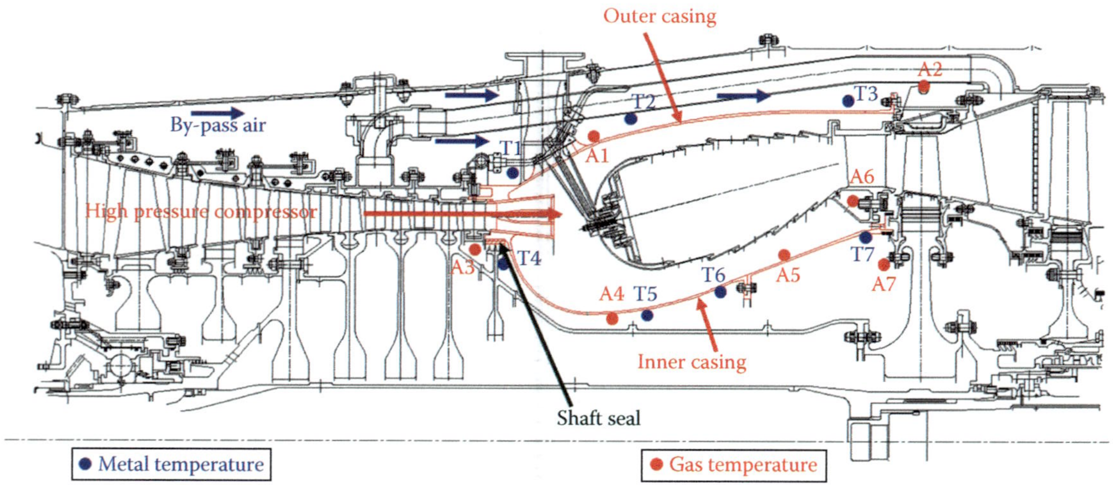

FIGURE 4.17
Location of thermocouples to measure combustor casing temperature. (From Panigrahi, S.K., Sarangi, N., and Chandrasekhar, U.: Experimental evaluation of overload capability of an annular combustor casing of a gas turbine engine. *J. Exp. Tech.* 2015. 40. 841–848. Copyright Wiley-VCH Verlag GmbH & Co. KGaA. Reproduced with permission.)

high-pressure cooling air supplied through the seal. The purpose of the seal is to allow a small, controlled flow of this cooling air to pass through the passage between the shaft and the liner. Both casings are exposed to the high-pressure compressor delivery air temperature ranging from 500 to 1000 K.

The mechanism of heat transfer to the combustor casing is by convection, radiation, and conduction. Flow passages are like annulus ducts. Heat transfer coefficients were estimated using the Dittus-Boelter equation. The next section presents the correlations that were used for the thermal analysis.

4.6.7 Nusselt Equation

Nusselt number (Nu) [137,138] is a dimensionless number and is the ratio of convective to conductive heat transfer across (normal to) the boundary.

$$Nu = \frac{(h \times D_{\mathrm{H}})}{k} \tag{4.1}$$

where

- Nu is the Nusselt number
- h is the convective heat transfer coefficient
- D_{H} is the annulus diameter, which is equal to the difference of the casing diameter and the combustor liner diameter
- k is the thermal conductivity of air

Typically, for forced convection, the Nusselt number is generally a function of the Reynolds number and the Prandtl number, or $Nu = f(Re, Pr)$. Empirical correlations for a wide variety of geometries are available that express the Nusselt number in the aforementioned forms. For the present type of flow, the Dittus Boelter equation was used.

4.6.8 Dittus Boelter Equation [139]

$$Nu = 0.0023 \times Re^{0.8} \times Pr^{0.4} \tag{4.2}$$

where

- Re is the Reynolds number (>2000)
- Pr is the Prandtl number (≈0.7)

From these two equations, the heat transfer coefficient is obtained as:

$$h = \left(0.023 \times Re^{0.8} \times Pr^{0.4}\right) \times \frac{k}{D_{\mathrm{H}}} \tag{4.3}$$

The heat transfer coefficients thus obtained were tabulated (see Table 4.3) together with the air temperature. The steady state nodal temperature was estimated using finite element analysis and was further validated against the measured metal temperature [140]. The estimated metal temperatures are shown in Figure 4.18. This nodal temperature was used in the finite element model to carry out the analysis.

Linear static analysis was carried out for the combination of pressure and thermal loads for the maximum operating condition.

The stress distribution over the combustor casing for the operating conditions at both pressure and thermal loads is given in Figure 4.19.

The ignitor boss region, where a crack is observed, experienced a hoop stress of 580 MPa, as shown in Figure 4.19. Therefore, for the present case, a submodel-based approach is preferred for detailed analysis of the stress concentration features, as high-quality and high-density mesh could easily be employed in a localized model. The submodel boundary deformations or tractions under the operating loads are obtained from the full three-dimensional model. The ignitor boss casing local zone, which was identified as a critical zone from a fatigue life point of view from the full-scale testing of the casing, is considered for further detailed numerical analysis.

TABLE 4.3

Heat Transfer Coefficient and the Estimated Air Temperature (see Figure 4.20)

Location	h (W/m² K)	Air Temperature (K)	Air Temperature (°C)
Zone 1A	181	519	246
Zone 1B	264	519	246
Zone 1C	270	519	246
Zone 2	110	896	623
Zone 3	789	896	623
Zone 4	1165	896	623
Zone 5	1705	896	623
Zone 6	1535	896	623
Zone 7	224	896	623
Zone 8	365	966	693
Zone 9	249	933	660
Zone 10	78	1016	743
Zone 11	2457	896	896

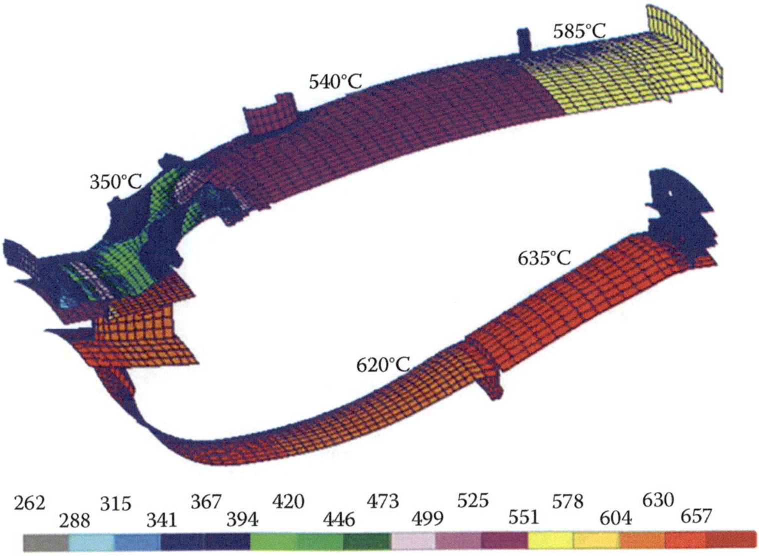

FIGURE 4.18
Metal temperatures at various areas of the casing in °C. (From Panigrahi, S.K., Sarangi, N., and Chandrasekhar, U.: Experimental evaluation of overload capability of an annular combustor casing of a gas turbine engine. *J. Exp. Tech.* 2015. 40. 841–848. Copyright Wiley-VCH Verlag GmbH & Co. KGaA. Reproduced with permission.)

4.7 Numerical Analysis of the Igniter Boss and Its Correlation with Experimental Data

Since the diameter of the combustor casing is quite large compared with the failed feature, a simplified model using a flat plate with a boss of 32 mm diameter was considered for the analysis. The submodel-based approach with a high-density mesh was preferred to the full-scale model for the detailed analysis of stress concentration features. The geometric details of the submodel of the ignitor boss area are shown in Figure 4.21. The solid model of the area is shown in Figure 4.22. The boss ignitor assembly is symmetric both in geometry and loading in the horizontal (x) and vertical (y) planes, so a finite element model of a quarter portion of the assembly is generated. The finite element model has been restrained in the x and y planes to suppress the rigid body motion mode and to simulate the full assembly. The full-scale component experiences a hoop stress of 580 MPa at the boss-shell area. Therefore, 580 MPa is applied as the load in the x-plane in the finite element model.

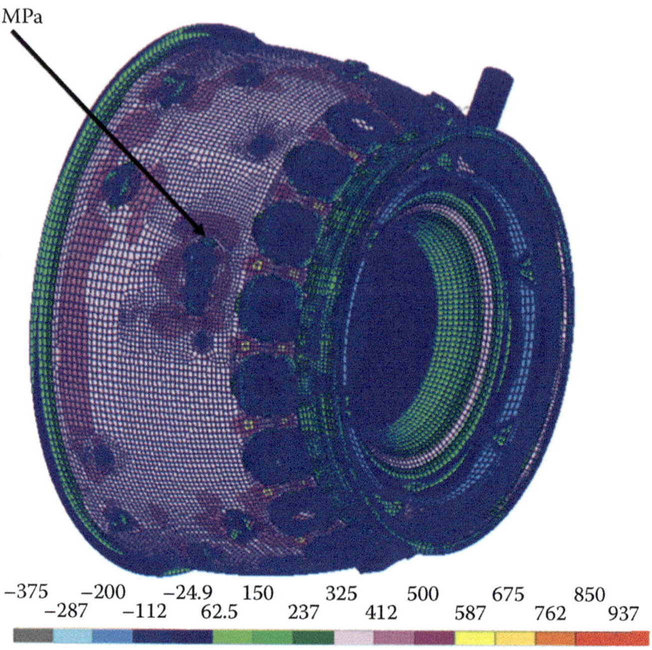

FIGURE 4.19
Hoop stress contour in the combustor casing (MPa). (From Panigrahi, S.K., Sarangi, N., and Chandrasekhar, U.: Experimental evaluation of overload capability of an annular combustor casing of a gas turbine engine. *J. Exp. Tech.* 2015. 40. 841–848. Copyright Wiley-VCH Verlag GmbH & Co. KGaA. Reproduced with permission.)

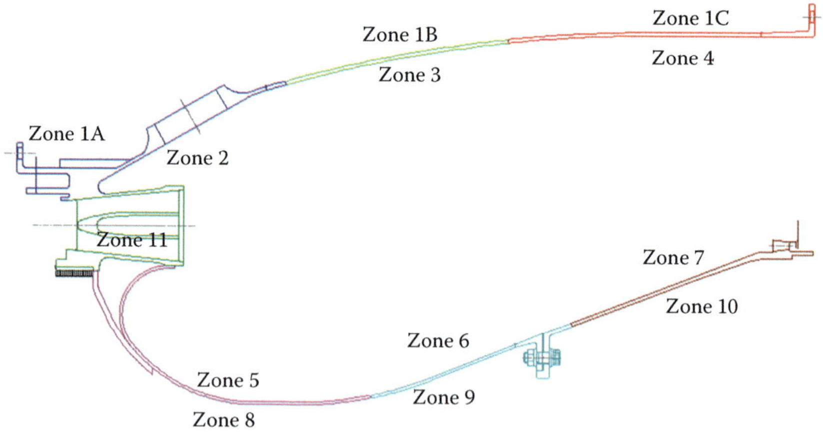

FIGURE 4.20
Air temperatures at various areas of the casing. (From Panigrahi, S.K., Sarangi, N., and Chandrasekhar, U.: Experimental evaluation of overload capability of an annular combustor casing of a gas turbine engine. *J. Exp. Tech.* 2015. 40. 841–848. Copyright Wiley-VCH Verlag GmbH & Co. KGaA. Reproduced with permission.)

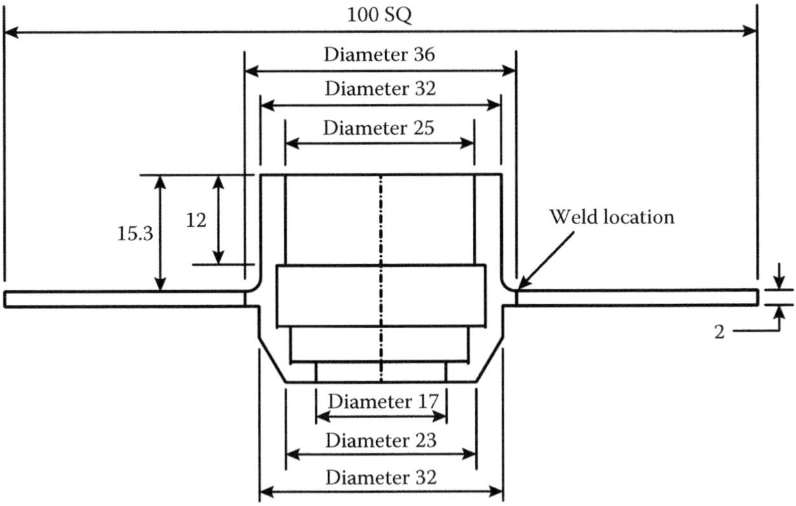

FIGURE 4.21
Boss-shell specimen showing the geometry and dimensions (all dimensions are in mm). (From Panigrahi, S.K., Sarangi, N., and Chandrasekhar, U.: Experimental evaluation of overload capability of an annular combustor casing of a gas turbine engine. *J. Exp. Tech.* 2015. 40. 841–848. Copyright Wiley-VCH Verlag GmbH & Co. KGaA. Reproduced with permission.)

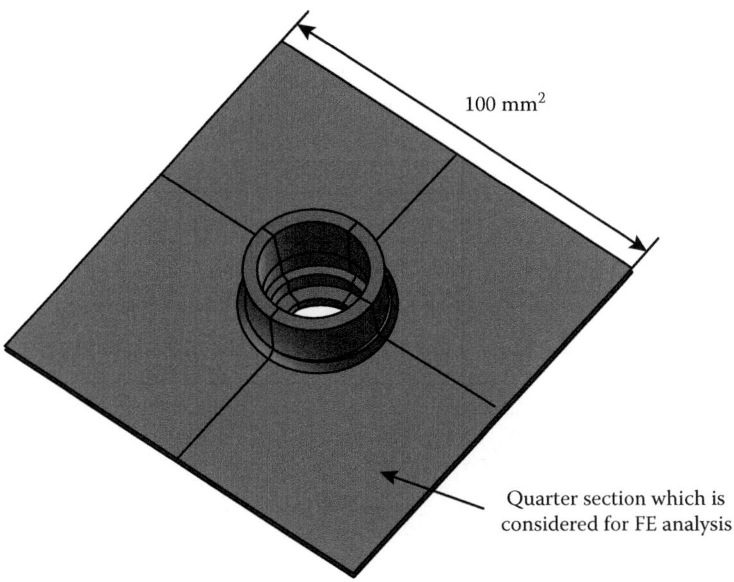

FIGURE 4.22
Solid model of the boss casing area. (From Panigrahi, S.K., Sarangi, N., and Chandrasekhar, U.: Experimental evaluation of overload capability of an annular combustor casing of a gas turbine engine. *J. Exp. Tech.* 2015. 40. 841–848. Copyright Wiley-VCH Verlag GmbH & Co. KGaA. Reproduced with permission.)

The finite element quarter model of the boss-shell zone is shown in Figures 4.23 and 4.24. A detailed finite element analysis of the critical zone at the ignitor boss in the casing was carried out and the low-cycle fatigue life of the casing was predicted using the Coffin–Manson approach. The finite element model of the boss ignitor area was developed using ANSYS mechanical analysis software with Solid45 elements.

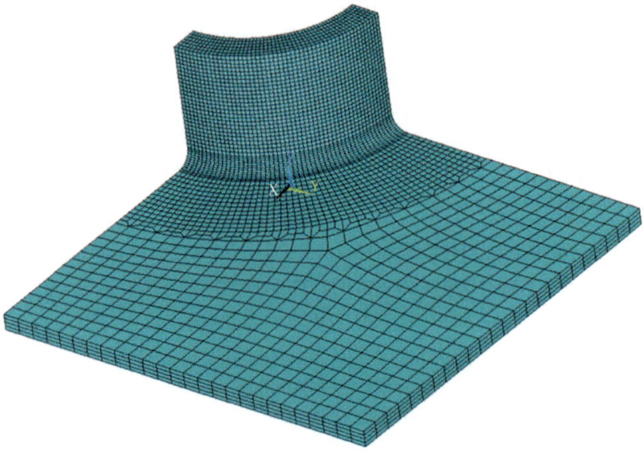

FIGURE 4.23
One quarter finite element model of the boss casing area. (From Panigrahi, S.K., Sarangi, N., and Chandrasekhar, U.: Experimental evaluation of overload capability of an annular combustor casing of a gas turbine engine. *J. Exp. Tech.* 2015. 40. 841–848. Copyright Wiley-VCH Verlag GmbH & Co. KGaA. Reproduced with permission.)

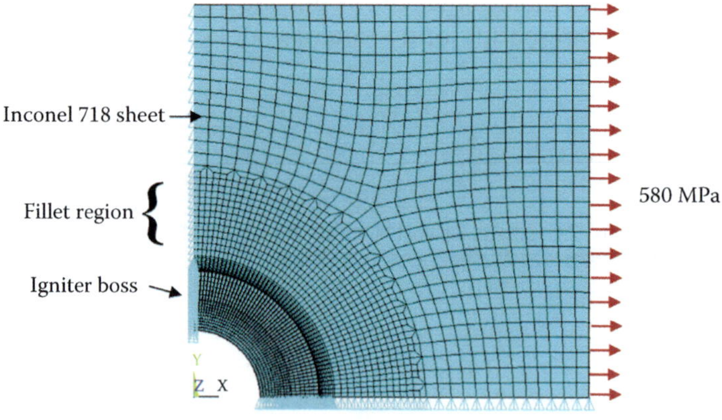

FIGURE 4.24
Boundary conditions applied in the boss casing area. (From Panigrahi, S.K., Sarangi, N., and Chandrasekhar, U.: Experimental evaluation of overload capability of an annular combustor casing of a gas turbine engine. *J. Exp. Tech.* 2015. 40. 841–848. Copyright Wiley-VCH Verlag GmbH & Co. KGaA. Reproduced with permission.)

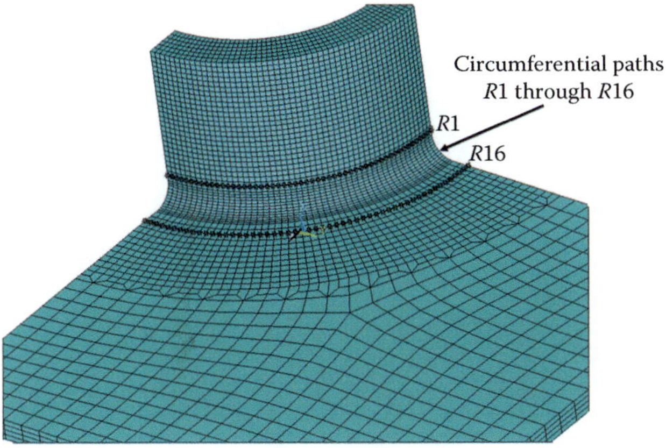

FIGURE 4.25
Path for results extraction. (From Panigrahi, S.K., Sarangi, N., and Chandrasekhar, U.: Experimental evaluation of overload capability of an annular combustor casing of a gas turbine engine. *J. Exp. Tech.* 2015. 40. 841–848. Copyright Wiley-VCH Verlag GmbH & Co. KGaA. Reproduced with permission.)

The finite element model consists of 26,368 elements and 32,394 nodes. Linear elastic stress analysis of the zone was carried out under the boundary traction equivalent to operating stresses from the global model of the combustor. The paths depicted in Figure 4.25 are used for results extraction. In the present study, the von Mises equivalent stress-based criteria [141–144] are selected as prediction models and the von Mises stress parameters in the simplified form were employed successfully to evaluate the multiaxial fatigue strength of the welded joint. The prediction parameters in this model are based on the local stresses obtained from the finite element simulated results [145–150].

The von Mises stress distribution in the boss ignitor zone is shown in Figure 4.26. The stress variation is plotted against the angle at various radii ($R1 = 16.01$ mm to $R16 = 20.5$ mm), as shown in Figure 4.27.

The maximum von Mises stress of 1080 MPa is observed at the end of the fillet at a radius of 17.491 mm and 950 MPa at the weld location at a radius of 18 mm. The presence of welds in the component affects the strength and life due to a variety of mechanisms as follows:

- Gas porosity
- Partial weld penetration
- Residual stress
- Heat-affected zone
- Mismatch of thickness at the weld joint

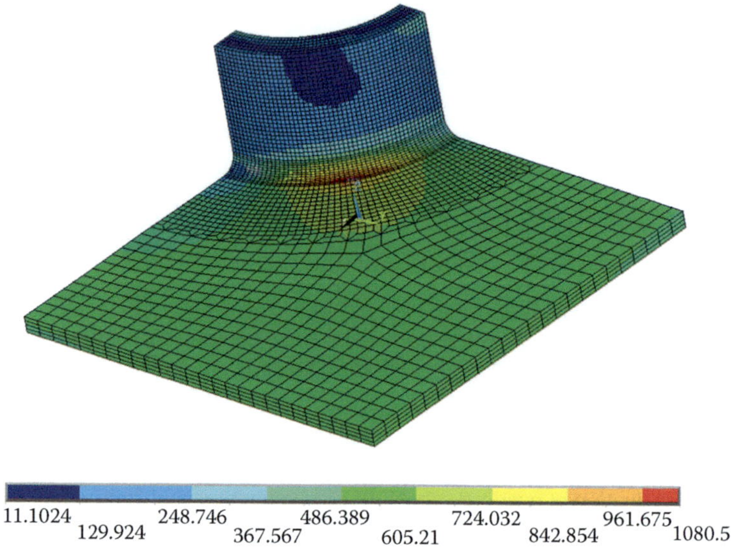

FIGURE 4.26
von Mises stress distribution in ignitor boss (in MPa). (From Panigrahi, S.K., Sarangi, N., and Chandrasekhar, U.: Experimental evaluation of overload capability of an annular combustor casing of a gas turbine engine. *J. Exp. Tech.* 2015. 40. 841–848. Copyright Wiley-VCH Verlag GmbH & Co. KGaA. Reproduced with permission.)

These factors affect the fatigue strength in welded structures to a great extent. Fatigue failures in welded joints happen mostly close to the welds, rather than in the base metal far from the weld.

Fatigue failure in the welded joints is a localized phenomenon and is affected or influenced by various parameters [151–155]. Since the weld is not modeled in the finite element analysis for the estimation of the fatigue life of the component at the location of the weld, a weld factor due to the weld, K_w, and a factor due to thickness mismatch, K_g, are employed for the estimation of fatigue life. The weld factor and the thickness mismatch factor are estimated based on analysis, the weld factor, K_w, is found to vary between 1.2 and 1.5, and the thickness mismatch factor, K_g, is found to vary in the range of 0.7–0.9.

Neuber's hypothesis [156,157] was used to arrive at the plastic strain experienced by the component. The Neuber method allows a conversion of fictitious wholly elastic stress values obtained from a finite element model to real elastic–plastic values. This is the basic condition for any computation done through the local elastic–plastic strain analysis (Figure 4.28).

The Coffin–Manson equation, as explained in Chapter 2, was used to establish the fatigue life of the component based on the mean stress effect. The constants in the Coffin–Manson equation are related to reduction in

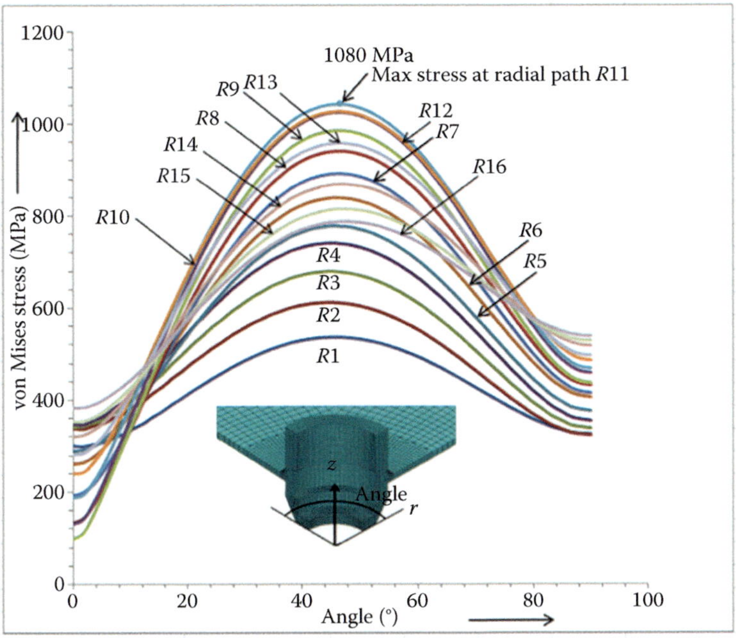

Radius	*R*1	*R*2	*R*3	*R*4	*R*5	*R*6	*R*7	*R*8	*R*9	*R*10	*R*11
In mm	16	16.01	16.039	16.128	16.266	16.412	16.586	16.784	17.004	17.241	17.491

*R*12	*R*13	*R*14	*R*15	*R*16
17.744	18	18.833	19.667	20.5

FIGURE 4.27
von Mises stress in the ignitor boss casing along the defined paths. (From Panigrahi, S.K., Sarangi, N., and Chandrasekhar, U.: Experimental evaluation of overload capability of an annular combustor casing of a gas turbine engine. *J. Exp. Tech.* 2015. 40. 841–848. Copyright Wiley-VCH Verlag GmbH & Co. KGaA. Reproduced with permission.)

area (RA), Young's modulus, and ultimate tensile strength (UTS). Since the load on the component is applied in terms of zero–maximum–zero, the component experiences alternating stress associated with a positive mean stress. The mean stress magnitude is equal to half of the stress amplitude. Therefore, mean stress correction is applied in the Coffin–Manson equation to evaluate the fatigue life of the component. The equation was rewritten (see Equation 4.4). As suggested by Manson and Halford [158,159], fatigue life can be estimated from the universal slope equation, as given here:

$$\nabla\varepsilon_t = D^{0.6}N_f^{-0.6} + \frac{3.5}{E}\left\{\text{UTS} - \sigma_{\text{mean}}\right\}N_f^{-0.12} \tag{4.4}$$

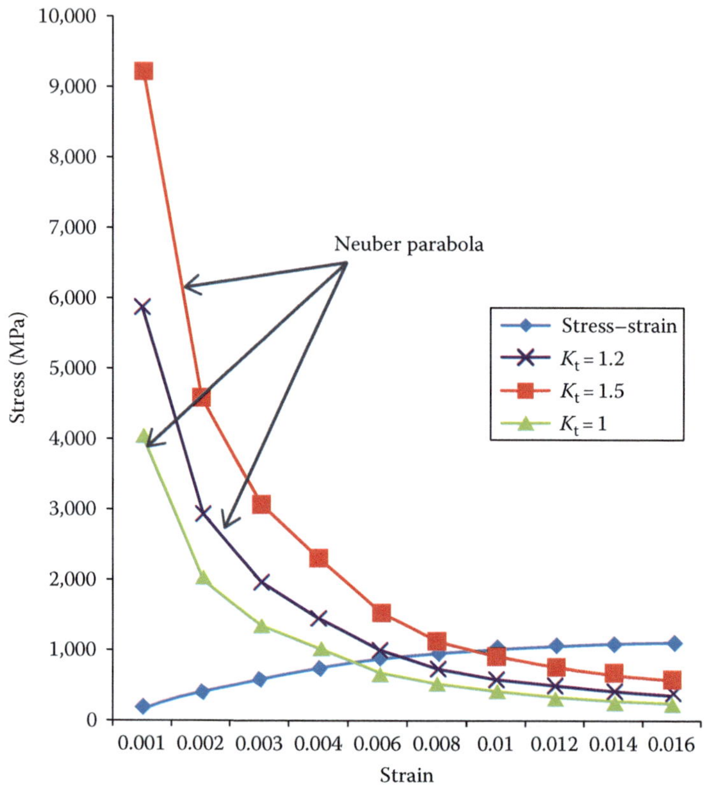

FIGURE 4.28
Neuber's methodology. (From Panigrahi, S.K., Sarangi, N., and Chandrasekhar, U.: Experimental evaluation of overload capability of an annular combustor casing of a gas turbine engine. J. Exp. Tech. 2015. 40. 841–848. Copyright Wiley-VCH Verlag GmbH & Co. KGaA. Reproduced with permission.)

A simple program is written in FORTRAN to compute the fatigue life, which is given here:

```
LCF PREDICTION
C     COFFIN-MANSON
C     WITH MEAN STRESS EFFECT
    Print*,"Input UTS IN Mpa"
    Read(*,*)UTS
    Print*,"Input E IN Mpa"
    Read(*,*)E
    print*,"Input RA IN %"
    Read(*,*)RA
    Print*,"Input Stress IN Mpa, FRACTION SIGMA_MEAN"
    Read(*,*)YP,SIGMA_MEAN
```

TABLE 4.4

Variation of Estimated Low Cycle Fatigue Life at Ignitor Zone with Various Parameters

S.N.	K_w	K_g	Nonlinear Peak Stress (MPa)	Elastic–Plastic Strain	Estimated Life
1.	—	—	800	0.005	22,830
2.	1.2	0.7	942	0.006	2,200
3.	1.2	0.8	942	0.006	3,170
4.	1.2	0.9	942	0.006	4,760
5.	1.2	1.0	942	0.006	7,440
6.	1.5	0.7	1033	0.01	990
7.	1.5	0.8	1033	0.017	1,290
8.	**1.5**	**0.9**	**1033**	**0.017**	**1,710**
9.	1.5	1.0	1033	0.017	2,380

```
   Print*,"Input epsi"
   Read(*,*)epsi
   SIGM= SIGMA_MEAN*YP
   D=-log(1.0-RA/100.0)
   cyn=100.0
   i=1
10   cyn=cyn+10.0
   i=i+1
   epsilon=3.5*((UTS-SIGM)/E)*CYN**(-0.12)+(D**0.6)
   *(CYN**(-0.6))
   print*,i,cyn,epsilon
     if((epsilon-epsi).gt.(0.0000001))go to 10
```

The factors K_w and K_g are incorporated in the equation. Considering the variation of these factors, LCF life at the ignitor boss was estimated and the details are given in Table 4.4. It can be noted from the table that the estimated fatigue life of the casing at the ignitor zone, without considering the life correlation factors, was found to be very high (22,830) compared with the fatigue life assessed by the full-scale testing of the combustor casing.

Based on this preliminary study, a combination of the parameters K_w and K_g was arrived at. The combination of K_w equal to 1.5 and K_g equal to 0.9 correlates well with the experimentally obtained life at the zone in the combustor casing. The sample calculation is as follows:

von Mises stress (σ) = 950 MPa

Stress concentration factor due to weld = K_w = 1.5

σ_{fic} = Linear stress= $\sigma \times K_w = 950 \times 1.5$ MPa = 1425 MPa

E = Modulus of elasticity = 221 GPa

Therefore, ε_{fic} = Linear strain = $\sigma_{fic}/E = 0.00645$

The Neuber constant is defined as the product of σ_{fic} and ε_{fic}, which is equal to 9.19.

From the Neuber curve (Figure 4.12), the real stress and strain are found to be 1033 MPa and 0.008, respectively.

UTS for the material = 1281 MPa

K_g = Thickness reduction factor = 0.9

$UTS_{corrected}$ = Corrected UTS = 1281 × 0.9 = 1153 MPa

R_A = Percentage reduction in cross-sectional area = 15 for the material

D = Ductility coefficient = ln(100/100 − RA) = 0.1625

σ_{mean} = Mean stress = 1033/2 = 516.5 MPa

Substituting all these values in the Manson and Halford equation (Equation 4.4), the estimated life (N_f) is computed as 1710 cycles.

4.8 Results and Discussions

The failure analysis of the failed zone attributed the failure to two main reasons. Firstly, a reduction in strength in the weld joint between the boss and the casing was due to a nonuniform weld. The nonuniformity in the weld occurred due to a variation in the wall thickness of the casing that had been manufactured from sheet metal by a forming process in order to avoid an axial weld on the casing. Secondly, the weld joint was located in a high-stress zone that developed around the boss due to interaction of the stress concentrations. Hence, it is suggested that the diameter of the collar of the boss be increased so that the weld joint is located away from the high-stress zone. The weld joint that was at the 36 mm diameter was moved to a 55 mm diameter by increasing the collar diameter, as shown in Figure 4.29. It was also suggested that, to achieve uniform wall thickness, the casing should be manufactured by machining from a forged billet instead of from sheet metal.

4.9 Summary

The fatigue life of a combustor casing was evaluated by full-scale testing in a simulated engine condition. The critical region in the casing was identified and a detailed analysis was carried out to predict the life of the casing.

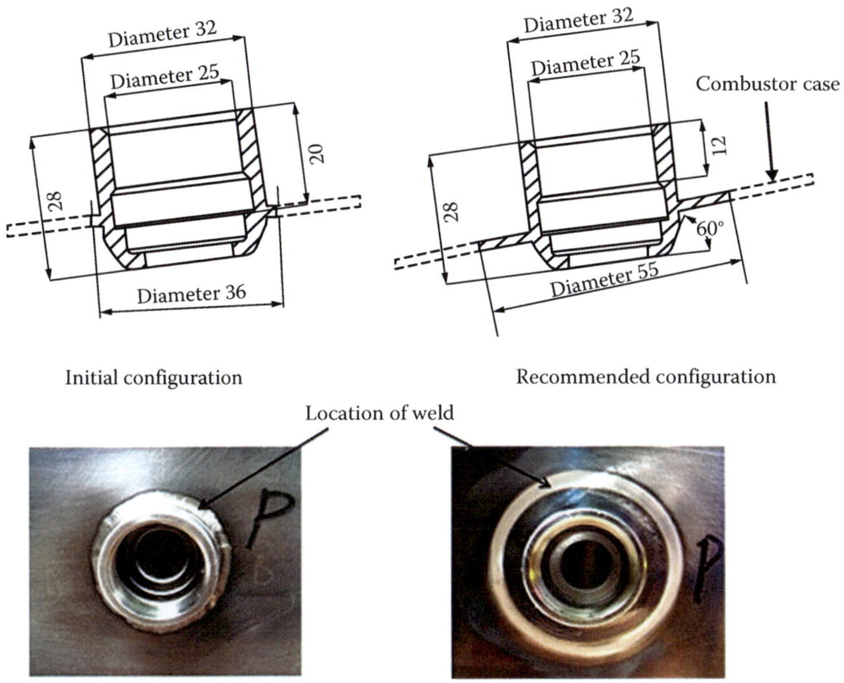

FIGURE 4.29
Configuration of the ignitor boss (all dimensions are in mm). (From Panigrahi, S.K., Sarangi, N., and Chandrasekhar, U.: Experimental evaluation of overload capability of an annular combustor casing of a gas turbine engine. *J. Exp. Tech.* 2015. 40. 841–848. Copyright Wiley-VCH Verlag GmbH & Co. KGaA. Reproduced with permission.)

Various factors which affect the fatigue life at the boss-shell weld in the casing were identified and considered for improved correlation of the casing life estimation. These factors have been refined based on the casing fatigue life obtained from the full-scale testing. The study provided useful input for improvement in the design and manufacturing processes in the combustor casing. The recommendations based on the study were incorporated in the design process and modified combustor casings were manufactured using the new methodology. The details of the modified manufacturing process and the reassessment of fatigue life of the modified combustor casing are explained in Chapter 5.

5

Reassessment of Fatigue Life of the Modified Combustor Casing

5.1 Introduction

In Chapter 4, we explained how cyclic pressure testing was conducted on a sheet metal combustor casing and that the test results were correlated with the numerical analysis carried out on a critical location. This testing and analysis provides very useful input for the improvement of the manufacturing process, weld quality control, and improved boss design. The means of component manufacture were changed to a forging process, which gave boss configuration that ensured the quality of weld, and afforded the thickness of the casing greater inherent strength and a longer life. In this chapter, we discuss the manufacturing methodology and, subsequently, the testing and validation of the modified combustor casing.

5.2 Modified Manufacturing Methodology

The combustor was machine forged to a thickness of 2 mm. This method has the advantage of better control of the thickness of the casing compared with the earlier method of construction from sheet metal. Details of the benefits of the new manufactured technique are described here.

5.2.1 Forging Process: Benefits and Drawbacks

Forging is the oldest known metalworking process. Forging was done historically by a smith using a hammer and anvil and, though the use of water power in the production and working of iron dates to the twelfth century CE, the hammer and anvil are not obsolete. The smithy has evolved over centuries into the forge shop with its use of engineered processes, production equipment, tooling, raw materials, and products to meet the demands of modern industry. In modern times, industrial forging is done either with presses or with hammers powered by compressed air, electricity, hydraulics, or steam. These hammers are large, having reciprocating weights in the thousands of pounds. Smaller power hammers, 500 lb or less reciprocating weight, and hydraulic presses are common in artists' smithies. Steam hammers are becoming obsolete.

Forging involves the shaping of metal using localized compressive forces. There are three kinds of forging processes: cold forging, which is done at room temperature or near room temperature; hot forging, which is done at a high temperature and makes metal easier to shape and less likely to fracture; and warm forging, which is done at an intermediate temperature. Forged parts can range in weight from less than 1 kg to 170 metric tons. Forged parts usually require further processing to achieve a finished part.

Forging results in metal that is stronger than that found in cast or machined metal parts. This stems from the grain flow caused through forging. As the metal is pounded, the grains deform to follow the shape of the part, thus the grains are unbroken throughout the part. Some modern parts take advantage of this for a high strength-to-weight ratio. Many metals are forged cold, but iron and its alloys are almost always forged hot. This is for two reasons: first, if work hardening were allowed to progress, hard materials such as iron and steel would become extremely difficult to work with; second, steel can be strengthened by other means than cold-working, thus it is more economical to hot forge than heat treat. Alloys that are amenable to precipitation hardening, such as most alloys of aluminium and titanium, can also be hot forged then hardened. Other materials must be strengthened by the forging process itself.

Hot forging is defined as working a metal above its recrystallization temperature. The main advantage of hot forging is that, as the metal is deformed, the strain-hardening effects are negated by the recrystallization process. Other advantages include the following:

- Decrease in yield strength, therefore it is easier to work and takes less energy (force).
- Increase in ductility.
- Elevated temperatures increase diffusion, which can remove or reduce chemical inhomogeneities.
- Pores may reduce in size or close completely during deformation.
- In steel, the weak, ductile, face-centered cubic (FCC) austenite is deformed instead of the strong body-centered cubic (BCC) ferrite at lower temperatures.

The disadvantages of hot working are

- Undesirable reactions between the metal and the surrounding atmosphere
- Less precise tolerances due to thermal contraction and warping from uneven cooling
- Grain structure may vary throughout the metal for a variety of reasons

Cold forging is defined as working a metal below its recrystallization temperature, but usually at around room temperature.

Cold forging has the following advantages:

- No heating required.
- Better surface finish.
- Superior dimensional control.
- Better reproducibility and interchangeability.
- Directional properties can be imparted into the metal.
- Contamination problems are minimized.

The disadvantages of cold working are as follows:

- Higher forces are required.
- Heavier and more powerful equipment and stronger tooling are required.
- Metal is less ductile.
- Metal surfaces must be clean and scale free.
- Intermediate anneals may be required to compensate for the loss of ductility that accompanies strain hardening.
- The imparted directional properties may be detrimental.
- Undesirable residual stress may be produced.

The manufacturing plan for the combustor casing is shown in Figures 5.1 and 5.2.

The three forgings were made separately out of Inconel-718 alloy, chosen to suit the combustor casing. Solid cylindrical billets were forged which were then converted to cylindrical hollow billets to suit the specified areas of the combustor casing. The different sizes of the three forgings are shown in Figures 5.1 and 5.2. The forgings were then machined to the required dimensions. This process had the advantage of machining both the casing

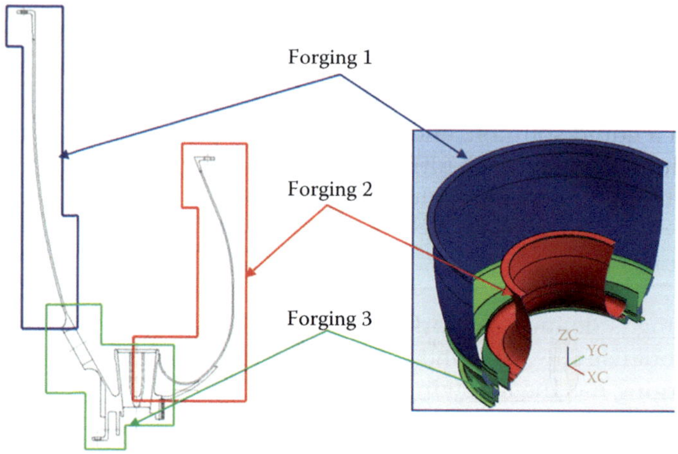

FIGURE 5.1
Manufacturing plan.

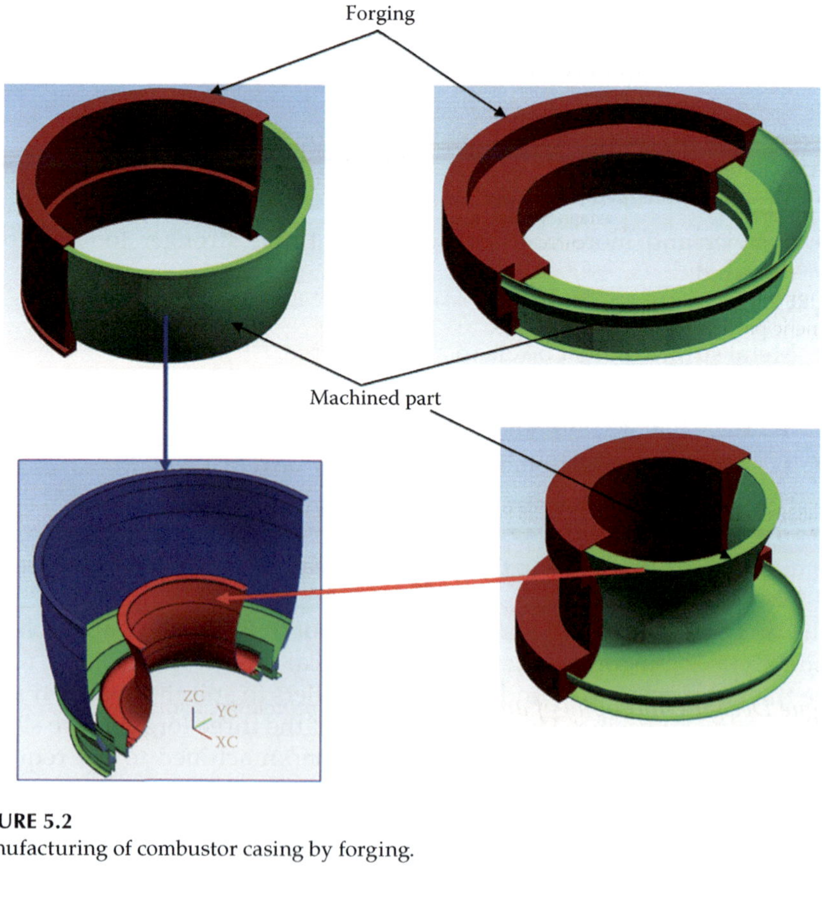

FIGURE 5.2
Manufacturing of combustor casing by forging.

and the flanges; as a result, the circumferential welding of the flanges at both the front and rear ends could be avoided. After machining, the parts were integrated with each other by electron beam welding (EBW). The holes required to mount the bosses were then drilled on the combustor casing. The bosses were machined from the forging bar stocks and were subsequently welded to the casing by tungsten inert gas (TIG) welding. Postweld heat treatment (PWHT) on the casing was carried out at various stages of welding to relieve the residual stresses induced during the welding process.

The following inspections were carried out on the casing to ensure the component was free from defects:

- Visual inspection
- Dimensional inspection as per the drawing and also at the critical location—the critical dimensions were taken to monitor them at various stages of the test
- Fluorescent particle inspection (FPI) over the entire surface
- Radiography at all the weld locations

The component was found to be in accordance with the design dimensions and was defect free. The modified forged combustor casing is shown in Figure 5.3.

After the manufacturing of the component, it is mandatory to conduct a pressure test on the component for the validation of its design and for the certification and validation of the new manufacturing methodology. The test setup described in Chapter 3 was used to conduct the test.

FIGURE 5.3
The forged combustor casing.

5.3 Instrumentation of the Modified Combustor Casing

Strain gauges (FLA-3-11, with gauge factor = 2.13) and strain rosettes (FCB-2-11, with gauge factor $X = 2.15$, $Y = 2.15$) are used to monitor strains at critical locations identified through stress analysis. The locations of the gauges are shown in Figure 5.4 and Table 5.1.

R1, R2, R3, R4, R5, R6, R7, and R8 are the two-dimensional rosettes (one dimension is circumferential and the other axial).

A, B, C, D, and E are the circumferential gauges (three gauges in each location with a gap of 7 mm between the gauges). The strain signals are required to monitor the behavior of the test component during the test.

The instrumented combustor casing is shown in Figure 5.5.

Pressure transducers (0–200 bar, accuracy ±1% of full scale), one in each area, were calibrated and used to monitor the pressure at the two areas.

5.4 Assembly and Trial Run

All the interface locations were sealed using O-rings made from neoprene rubber; these O-rings had diameters of 2.5 or 3.5 mm according to the areas being sealed. Teflon sheets were used to seal the flat surfaces at the ignitor. Bonded seals were used in the blanking areas to prevent oil leakage during testing. All the adaptation components, manufactured as discussed in Chapter 3, were used to adapt the combustor casing in the test facility. The strain gauged combustor was assembled with leak-proof sealing plates and other structural elements in the dual channel servo hydraulic pressure test facility, as shown in Figure 5.6.

FIGURE 5.4
Location of strain gauges.

TABLE 5.1

Location of Gauges in Modified Combustor Casing

Sl. No.	Gauge Name/No.	Axial Distance in mm from the Front Flange	Angle Measured (°) Clockwise When Viewed from the Front Flange with Reference to the Reference Line (Figure 5.4)	Remarks
1	R1	440	0	Smooth area
2	R2	365	0	Smooth area
3	R3	290	0	Smooth area
4	R4	220	0	Smooth area
5	R5	140	0	Smooth area
6	R6	100	0	Smooth area
7	R7	50	0	Smooth area
8	R8	10	0	Smooth area
9	A1	360	50	Boss housing drain valve
10	A2	353	50	
11	A3	346	50	
12	B1	165	50	Boss for total pressure-1
13	B2	158	50	
14	B3	151	50	
15	C1	335	135	Boss for total pressure-2
16	C2	328	135	
17	C3	321	135	
18	D1	205	135	Light up atomizer Boss
19	D2	205	135	
20	D3	205	135	
21	E1	180	120	Boss igniter
22	E2	173	120	
23	E3	166	120	

The test component was filled with hydraulic oil through the inlet ports, keeping the air vents open. The air vents were closed after zones A and B were completely filled. The zones were individually pressurized to ensure there were no leaks between the zones or at the boss locations. Several test cycles were carried out at low pressure (15 bar) to ensure the test setup was ready to conduct the cyclic pressure test on the combustor casing in accordance with the program requirement. The test setup and the component inspection reports were reviewed by a committee of certifying authorities and approval was given to conduct the test.

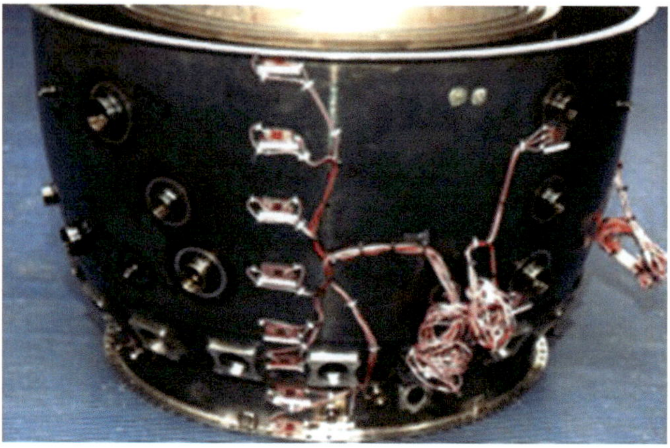

FIGURE 5.5
Instrumented combustor casing.

FIGURE 5.6
The new combustor casing during testing.

5.5 Testing

The component was subjected to the following test cycle to demonstrate its performance for 7000 numbers of fatigue cycles.

Base pressure (bar): Atmospheric

Peak pressure (bar): 29.8—Main chamber (zone A) and 19—Inner chamber (zone B)

Test temperature: Ambient

Pressurization time(s): 20

Dwell time(s): 6 (at peak pressure)

Depressurization time(s): 20

Total no. of cycles: 7000 cycles

The test cycle for the cyclic pressure test is shown in Figure 5.7.

5.6 Results and Discussion

Conducting a fatigue life test on a full-scale combustor casing is a challenging task. The test setup described in Chapter 3 was used to conduct the test. The modified combustor casing was subjected to the pressure profile as indicated earlier for 7000 nm cycles, to meet the mandatory airworthiness requirement. The component was inspected dimensionally, by FPI and by

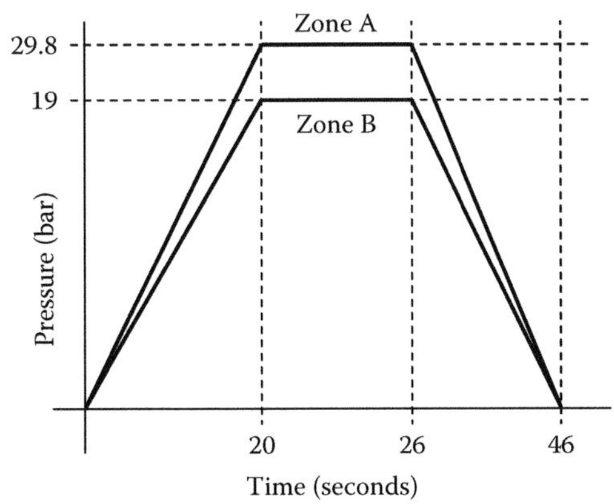

FIGURE 5.7
Test cycle for cyclic pressure test on new combustor casing.

TABLE 5.2

Comparison of Version 1 of the Combustor Casing with the Modified Combustor Casing

Sl. No.	Parameter	Version 1 of Combustor Casing	Modified Version of Combustor Casing
1	Manufacturing process	Metal spinning	Forged
2	Wall thickness of 2 mm	Variation observed from 1.6 to 2.2 mm	Precise thickness could be maintained
3	Boss geometrical features	Base diameters of the bosses were close to the high stress zone	Base diameters of the bosses were increased and were taken away from the high stress zone
4	Number of circumferential welds	Two in the outer casing and two in the inner casing	One in the outer casing and one in the inner casing
5	Cost and time required for manufacturing	Economical and faster	More costly and more time required
6	Fatigue life	Shorter	Extended

radiography, before starting the test and at intervals of 2000, 5000, and 7000 cycles. The FPI reports observed no surface cracks, radiography inspection revealed there were no cracks at weld locations and the dimensional inspection observed there were no changes in dimensions. To establish sufficient reliability on the combustor casing based on the test conducted on a single component, it was proposed to subject the test component to 28,000 cycles with stage inspections at 14,000, 21,000, and 28,000 cycles. The combustor casing withstood 28,000 numbers of minimum–maximum cycles and demonstrated its fatigue life performance. The range of tests successfully demonstrated the manufacturing methodology recommended. The comparison of the modified version of the combustor casing with the first version is summarized in Table 5.2.

Though the component demonstrated its fatigue life, it is a mandatory requirement to test a component from the perspective of its safety. To comply with airworthiness requirements, a component has to be subjected to twice the maximum operating pressure for a duration of 300 s in order to be certified from the point of view of safety.

5.7 Summary

The recommendations based on the test results of the sheet metal combustor casing were implemented and a new combustor casing was manufactured by forging. The modified combustor casing was subjected to fatigue loading in the pressure test facility to demonstrate its fatigue life. The modified combustor casing successfully demonstrated 28,000 cycles without any deformation. A new modified combustor casing was considered for conducting proof pressure tests; the details of the test are explained in Chapter 6.

6

Safety Test on Modified Combustor Casing

6.1 Introduction

A combustor casing can be considered to be a pressure vessel designed to withstand hot gases at substantial pressures up to 30 bar. The pressure differential is dangerous and fatal accidents may occur during operation. Pressure vessels hold a considerable amount of energy when at working pressure and therefore it would be disastrous if they burst in operation. Hence, the problem of pressure vessels failing has been under constant and extensive research [160]. In recent years, in line with design criterion analysis of important and critical components of machinery, aero engines have been required to undergo the evaluation of the burst capability of relevant components or pressure vessels. Accurate calculation of the bursting pressure can improve not only safety factors but also design precision, which can lead to a decrease in the quantity of materials used in manufacture, thereby lightening the weight of pressure vessels.

The design of pressure vessels for critical high-pressure applications has to consider two modes of failure [161,162]. The first mode of failure can occur when the deformation in operation becomes excessive and permanent deformation occurs. The second type of failure occurs at higher magnitudes of pressure resulting in the pressure vessel bursting and catastrophic failure. Hence, predicting the burst pressure or proof pressure capability of a pressure vessel is an important design consideration as it gives a more realistic indication of the safety margin over the maximum expected operating pressure.

Consequently, combustor casing design, manufacture, and operation of aero engines are regulated by certifying authorities. The strength of the combustor casing is usually tested by measuring the permanent deformation of the casing after subjecting the casing to the rated pressure. Hydrostatic testing is the most common method employed for testing aero engine pressurising components. Using this test helps to maintain safety standards and the durability of the component over time. Newly manufactured pieces are initially qualified using the hydrostatic test. They are then requalified for safety using the proof pressure test, where pressure of the order of twice the maximum operating pressure is applied for a specific duration to ensure the combustor casing's safety capability. Testing of such components is very important because they can explode if they fail under pressure, causing catastrophic damage to the aircraft.

A proof pressure test is a form of stress test to demonstrate the fitness of a pressure-loaded structural component of an aero engine. An individual proof test may apply only to the unit tested, or to its design in general for mass-produced items. Such a structure is often subjected to loads above that expected in actual use, demonstrating the safety and design margins. Proof testing is nominally a destructive test in case of an aero engine application. Proof tests are performed before a new design or unit is allowed to enter into service, or to perform additional uses, or to verify that an existing unit is still functional as intended. All pressure vessels/systems must be certified as safe to operate from a pressure viewpoint before use.

"Leak before burst" describes a pressure vessel designed such that a crack in the vessel will grow through the wall, allowing the contained fluid to escape and reducing the pressure, prior to growing so large that it causes fracturing at the operating pressure. Many pressure vessel standards, including the American Society of Mechanical Engineers (ASME) Boiler and Pressure Vessel Code and the American Institute of Aeronautics and Astronautics (AIAA) Metallic Pressure Vessel Standard, either require pressure vessel designs to be "leak before burst," or require pressure vessels to meet more stringent requirements for fatigue and fracture if they are not shown to be "leak before burst."

The modified combustor casing, which was manufactured by machining out of forged billets, demonstrated an adequate fatigue life. However, the component needs to be declared fit to fly from the safety point of view, for which it must withstand twice the maximum operating pressure for a duration of 300 seconds. The test must be conducted on a new component; therefore, a new modified combustor casing was prepared for conducting the proof pressure test to demonstrate its proof capability. This chapter presents the results of the proof pressure test carried out on the modified combustor casing of an aero engine to indicate its safety so as to obtain airworthiness certification of the aero engine.

The proof strength of the combustor casing directly affects the safety of the engine, as well as the aircraft. Therefore, it is mandatory to ensure the strength of the casing. Major factors that determine the proof pressure of the casing are stress concentration around the surface and the welding of the bosses, as well as the casing's shape and thickness.

6.2 Test Component Details

The configuration of the component is same as explained in Chapter 5 and is shown in Figure 6.1.

FIGURE 6.1
The forged combustor casing identified for the proof pressure test.

The component was inspected by dye penetrant inspection, by radiography inspection to confirm the component as free from defects, and by dimensional inspection carried out to establish the component dimensions before starting the test.

6.3 Instrumentation and Testing

Strain gauge rosettes are mounted on the component at critical locations identified by analysis, as shown in Figure 6.2.

The test set up and the test facility described in Chapter 5 was used to conduct the proof pressure test in accordance with the following test cycle:

Base pressure (bar): Atmospheric

Peak pressure (bar): 57—Main chamber (zone A) and 36—Inner chamber (zone B)

Test temperature: Ambient

Pressurization time(s): 20

Dwell time(s): 300 (at peak pressure)

Depressurization time(s): 20

Total no. of cycles: 1

The combustor casing during the test is shown in Figure 6.3.

FIGURE 6.2
The combustor casing with strain gauges prior to proof pressure test.

FIGURE 6.3
The combustor casing during proof pressure test.

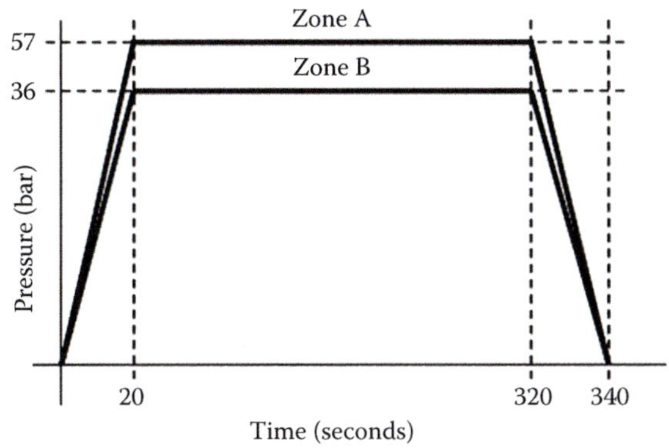

FIGURE 6.4
The proof pressure test profile.

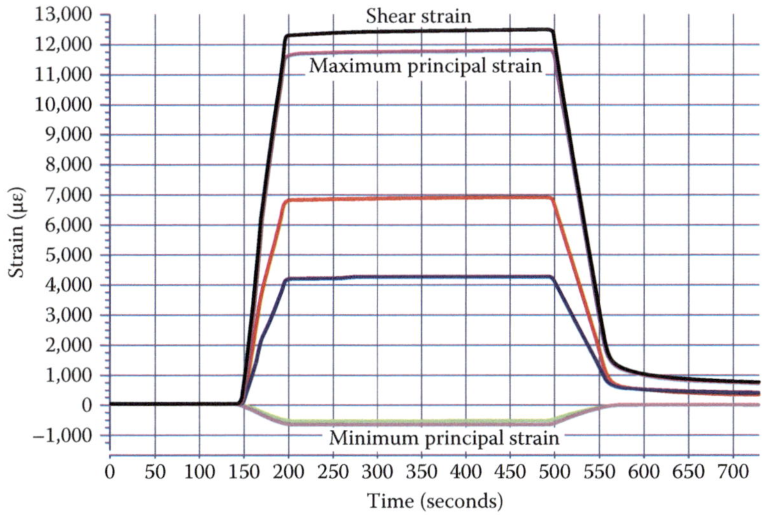

FIGURE 6.5
The strain signals during proof pressure test.

Pressure values and the strain values are monitored during the test. The pressure cycle is shown in Figure 6.4 and strain cycles captured during the test are shown in Figure 6.5.

6.4 Results and Discussion

The component under test experiences maximum load during the proof pressure test and conducting a proof test on a full-scale combustor casing is a challenging task. The component was assembled in the test setup in accordance with the engine configuration. Fluorescent particle inspection (FPI), radiography at the weld locations, and a dimensional inspection were carried out before starting the test and at the end of the test.

No surface cracks were noted in the FPI reports, no cracks were found at weld locations during the radiography inspection, and no change to the dimensions were observed in the dimensional inspection report. The combustor casing withstood the proof pressure test successfully and demonstrated its fitness for the performance of a safe life.

6.5 Summary

The proof pressure test was conducted on a new combustor casing which met the safety requirement of the combustor. As was discussed in Chapter 5, the cyclic pressure test was conducted on a different component than the one that met the life requirement of the combustor. However, the objective of the structural design of an aero engine combustor is primarily to provide the engine combustor with sufficient fatigue life and strength to permit its continuous operation until a scheduled engine overhaul and to withstand certain overload. During operation, a combustor is subjected to cyclic pressure which may affect its over pressure capability. Therefore, the proof pressure capability of a combustor casing which has already subjected to certain amount of fatigue loading needs to be established. The details of such a test are explained in Chapter 7.

7

*Effect of Fatigue on the Proof Strength of an Aero Engine Combustor Casing**

7.1 Introduction

The objective of an aero engine combustor structural design is primarily to provide an engine combustor with sufficient fatigue life and strength to permit its continuous operation between scheduled engine overhauls, and to withstand a level of operational overload. During operation, a combustor is subjected to cyclic pressure, which may affect its overpressure capability. The combustor casing is considered to be a critical part in an engine. In this chapter, we evaluate the proof pressure capability of a combustor casing subjected to fatigue loading. We also establish the effects of fatigue damage on the proof pressure capability of the combustor casing. During the testing of the casing, the pressure was continuously increased to evaluate the overpressure capability of the casing. The results of this experimental study on the casing of an aero engine combustor are presented in this chapter. A detailed post-fracture investigation of the fracture surfaces of the casing shows that two independent cracks developed: one along the circumference of the casing towards the front flange and the other in an axial direction. Further detailed investigation of the fractured surfaces under a stereo zoom microscope (SZM) and scanning electron microscope (SEM) shows that the presence of striations, coupled with dimples, indicates that the failure was due to fatigue loading. Details of the metallographic techniques are described in the following section.

* Reprinted with permission from Sarangi, N., Panigrahi, S.K., Patel, S.K., Ramesh, P., Ganeshachar, M.D. Effect of fatigue on proof strength of gas turbine combustor casing, *J. Fail. Anal. Prev.*, 13(1), 2013, 102–111.

7.2 Metallographic Techniques in Failure Analysis

Metallographic examination [162] is one of the most important procedures used by metallurgists in failure analysis. It is also a prime tool in metallurgy to characterize the microstructure of the materials. Analysis of the microstructure of a material aids in determining if the material has processed correctly; such analysis is therefore a critical step for determining product reliability and for determining why a material or a component has failed. Metallographic analysis is a valuable tool. By properly documenting the initial condition of a sample or the fractured surface of a failed component and the proceeding microstructural analysis, metallography provides a powerful means of quality control and an invaluable investigative tool. Metallurgical analysis (metallography) of the microstructure of a sample provides the scientist or metallurgist information varying from phase structure, grain size, solidification structure, casting voids, and so on. The basic steps in the proper preparation of a metallographic sample include documentation, sectioning and cutting, mounting, planar grinding, rough polishing, final polishing, etching, microscopic analysis, and hardness testing.

Development of powerful electron metallographic instruments, such as the SZM and SEM, has not diminished the importance of optical microscopy. Basically, the optical microscope is used to assess the nature of the microstructure of a sample and its influence on the failure mechanism. The purpose of using an optical microscope is twofold: it can reveal the relationship between the microstructure of a sample and the crack path (in failures involving fracture), and/or the nature of corrosion or wear damage; it can also be determined whether processing or service conditions have produced undesirable microstructural conditions that have contributed to failure, such as abnormalities due to material quality, fabrication, heat treatment, and operational conditions.

Examination of fractured components should begin with a low-power stereo microscope. Handheld magnifying lenses are still widely used to study fractures but this mainly takes place in the field. While the optical microscope has limited value for direct observation of fracture surfaces, a great deal can be learned by indirect examination, that is, by examination of the fracture profile and secondary cracking. Detailed observation of the fracture surface is best accomplished with use of an SEM. However, lack of access to an SEM should not be viewed as a crippling obstacle to performing failure analysis, because such work was undertaken successfully prior to the development of these instruments. In many studies, such equipment is not needed; in other cases, SEMs are very important tools as, in most cases, a more thorough job can be accomplished using such tools.

7.2.1 Stereo Zoom Microscopy

Stereo microscopes and zoom microscopes both generate three-dimensional and laterally precise images for metallographic applications. These instruments offer large object fields and allow extended working distances. A stereo microscope is perfect for the observation of large biological samples such as leaves or tissues. In the field of microscopic materials, stereo and zoom microscopes are valuable tools for the inspection or analysis of the surfaces of rough materials.

An SZM is an optical microscope that functions at low magnification. It works by using two separate optical paths instead of just one. The two objectives and two eyepieces provide slightly different viewing angles. In essence, the left and right eye are seeing the same object but each in a different way. In much the same way as happens with our actual eyes, these two separate viewing angles yield a three-dimensional image. This feature makes an SZM ideal for examining the surfaces of solid materials. SZMs also lend themselves well to sorting and dissecting. One can use these units for working with watches, circuits, and even microsurgery.

The lighting in an SZM—reflected illumination—is also different than that of other types of microscopes. This means that it utilizes light that is naturally reflected from the object. This is ideal when dealing with thick or opaque samples. There are some models of SZM that can use transmitted light. The bulb or mirror is located beneath the object itself. However, the lighting is not controlled through a condenser, as is the case on compound microscopes. Certain stereo models can also be used in dark field microscopy when necessary. An SZM provides a three-dimensional view of the sample. It does this with separate objective lenses and eyepieces for each eye. They have lower magnification when compared with compound microscopes, but they also have a longer working distance.

Recently, CCD camera pickups have been fitted to stereo microscopes, allowing the images to be displayed on a high resolution LCD monitor. Software converts the two images into an integrated 3D image, for viewing with plastic red/cyan glasses, or to the cross-converged process for clear glasses and somewhat better color accuracy. This permits the results to be viewed by groups of people, rather than just the analyst. More typically, a camera attached to one of the eyepieces will record conventional 2D images. A typical stereo microscope is shown in Figure 7.1.

7.2.2 Scanning Electron Microscopy

The SEM is one of the most versatile instruments for investigating the microstructure of materials. Under electron bombardment, a variety of different signals is generated (including secondary electrons, backscattered electrons, characteristic x-rays, and long-wave radiation in the

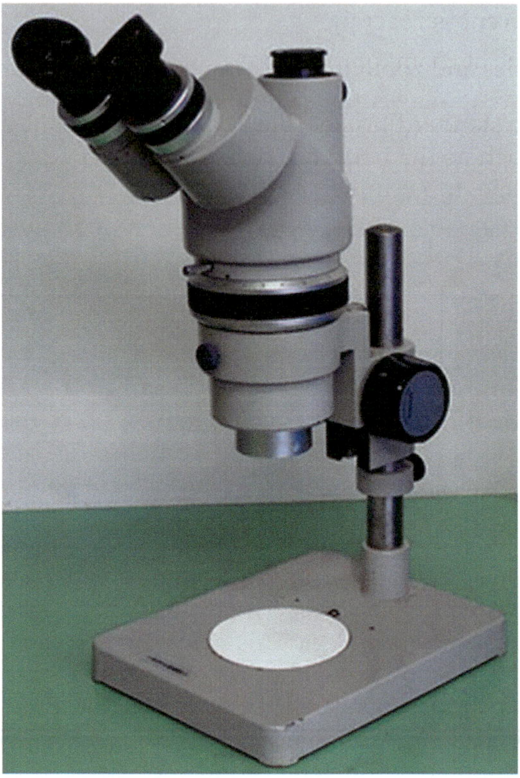

FIGURE 7.1
A stereo microscope. (With kind permission from Springer Science+Business Media: *J. Fail. Anal. Prev.*, Effect of fatigue on proof strength of gas turbine combustor casing, 13(1), 2013, 102–111, Sarangi, N., Panigrahi, S.K., Patel, S.K., Ramesh, P., Ganeshachar, M.D.)

ultraviolet and visible region of the spectrum) that can be used for the characterization of materials. Using secondary electrons, SEM expands the resolution range to a few nanometers (nm) (under favorable conditions), thus bridging the gap between optical (light) microscopy and TEM. In addition to the higher lateral resolution, SEM also has a much greater depth of field compared with optical microscopy, due to the small size of the final lens aperture and the small working distance. SEM offers possibilities for the formation of images that are usually easy to interpret and will reveal clear pictures of polished and etched cross-sections as well as rough surfaces and particles. Energy-dispersive x-ray microanalysis using equipment routinely attached to the SEM features qualitative and quantitative analysis of the chemical composition of a sample with a typical lateral resolution of 1 μm and a typical depth resolution of a few tenths of a micrometer. Due to its relatively easy handling, SEM has found a wide

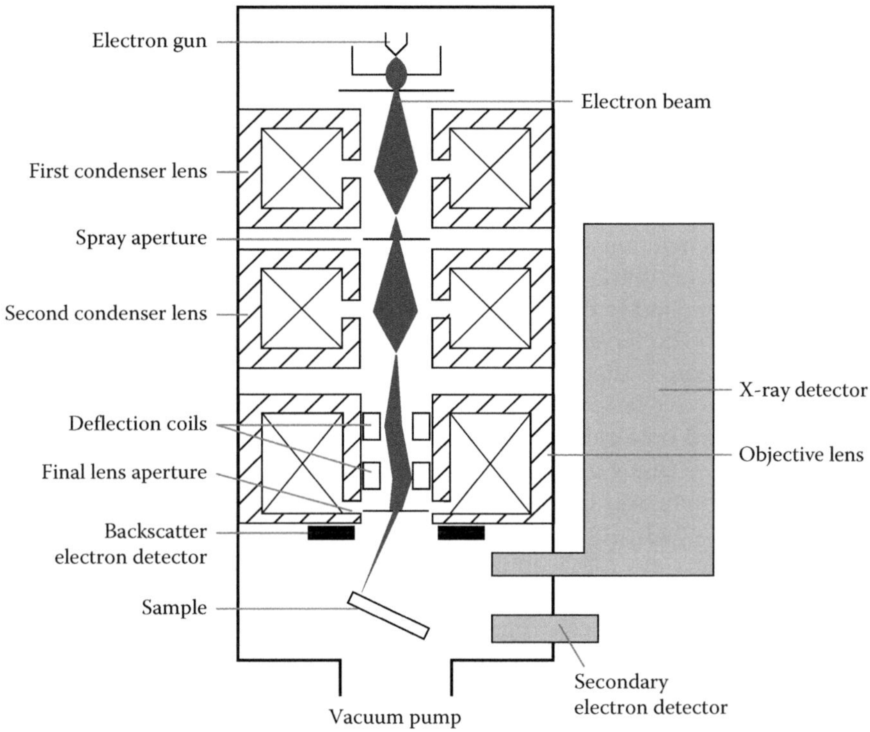

FIGURE 7.2
A schematic of a scanning electron microscope. (Courtesy of Wikipedia, "Scanning electron microscope," https://en.wikipedia.org/wiki/Scanning_electron_microscope#/media/File:Schema_MEB_(en).svg.)

range of applications in materials research, materials development, failure analysis, and quality control.

An SEM is a microscope that uses electrons, rather than light, to create a magnified image and can magnify well above 100,000 times (3 nm resolution). Optical microscopes are only able to magnify approximately 1000 times. An SEM can be used to look at a variety of materials, such as minerals and manufactured materials. The main components of an SEM include the electron gun, probe-forming column (consisting of magnetic electron lenses, apertures, and scanning coils), electron detectors, and a vacuum system, as shown in Figure 7.2.

In the electron gun, electrons are generated by thermionic emission or by an electric field. These electrons are then accelerated towards the sample. The most common type of electron gun consists of a tungsten filament that acts as a cathode. The anode consists of a grounded plate with a hole through which the electrons pass. The accelerating voltage is usually varied between

1 and 50 kV. A third electrode (Wehnelt cylinder) with a negative bias of a few hundred volts (relative to the cathode) is introduced to limit the emitting area to the tip of the tungsten filament. Lanthanum hexaboride (LaB_6) cathodes are used in order to obtain a higher brightness; this has additional benefits, such as improving the lateral resolution. Field emission sources with even higher levels of brightness than the LaB_6 cathode recently became available for routine operation in commercial instruments.

Magnetic electron lenses are used to demagnify the image of the electron source (cross-over) to the final spot size on the sample surface. In most cases, three lenses are used to obtain the required demagnification. Apertures are placed between the lenses to limit the beam diameter. In order to scan the electron beam across the sample, two sets of scanning coils are placed in the bore of the objective lens. The magnification is varied by changing the size of the area scanned on a sample. The electron beam scans the sample in much the same way as in a cathode ray tube (CRT) used for image formation on a television screen, and the output of the electron detectors is displayed on the screen of a synchronously scanned CRT. In modern instruments, analog scanning systems are replaced by digital systems in which the movements of the beam on a sample are controlled with a computer. The analog signal from the electron detectors is digitized and stored as a number for each pixel.

The electron gun and the column must be evacuated in order to avoid damage to the electron source and high-voltage breakdown in the gun. A high vacuum is also necessary to minimize scattering of the electrons during their travel from the gun to the sample surface. In general, the vacuum system consists of a high-vacuum pump (oil-diffusion pump, turbo molecular pump, or ion pump) and a mechanical rotary pump. Operation of a tungsten filament requires a vacuum in the gun of better than $\approx 10^{-3}$ Pa; an LaB_6 cathode, better than $\approx 10^{-5}$ Pa; and a thermal field emitter, better than $\approx 10^{-7}$ Pa. The high vacuum in the instrument leads to evaporation of volatile compounds, especially under electron bombardment, and sample characterization may be limited by the stability of the material. Recently, environmental SEM has become a routine technique. In these instruments, pressures up to 2600 Pa are possible in the sample chamber during imaging with secondary electrons (SEs) and backscattered electrons (BEs). Due to multiple pressure-limiting apertures, a high vacuum can be retained simultaneously in the electron gun and the column. The relatively high pressures in the sample chamber enable the study of fragile materials, such as biological tissue, plastics, and grease. It is even possible to stabilize liquid water during electron microscopical investigation.

The electron detector most commonly used in SEM is the Everhart–Thornley detector, which consists of a scintillator that, under electron bombardment, produces photons. The photons are converted to an electrical signal by means of a photomultiplier. The Everhart–Thornley detector can be used for SEs and BEs. However, there are more dedicated BE detectors, including large-area scintillator detectors, BE-to-SE conversion detectors, or solid-state diode detectors. The latter may be divided into sectors in order to obtain different contrast by combining the output of the individual sectors in different ways (see the following). In environmental SEM, it is necessary to use special gaseous SE and BE detectors, which use gas ionization to detect and amplify the signal.

In addition to the main components of an SEM, described previously, most instruments are equipped with an energy dispersive x-ray detector (silicon or germanium solid-state detector). Energy dispersive x-ray spectroscopy enables the qualitative and quantitative chemical analysis of elements with an atomic number ≥5 (boron). X-ray detection may also be carried out by wavelength-dispersive spectrometers, which consist of a crystal that reflects radiation of a specific wavelength (Bragg's law) and a proportional counter for x-ray detection. Additional detectors that may be fitted to an SEM include a cathode luminescence detector (photomultiplier) and acoustic or infrared detectors. No special detector is needed for registration of the absorbed current, because the sample itself acts as a detector.

7.2.3 Optical Microscopy

The optical microscope remains the most important tool for the study of microstructure, despite the evolution of sophisticated electron metallographic instruments. SEM and transmission electron microscopy (TEM) are also invaluable tools; however, they should be used in conjunction with optical microscopy, rather than as a substitute. For more information on these methods, see Section 7.5.3. For the efficient assessment of the basic characteristics of a microstructure, all examinations of microstructure should begin with use of an optical microscope, starting at low magnification, such as 100×, followed by progressively higher magnifications. Most microstructures can be observed with an optical microscope and identified based on their characteristics. Identification of questionable or unknown constituents may be aided by observation of their hardness relative to the matrix, by their natural color, by their response to polarized light, and by their response to selective etchants. These observations are compared with known details about the physical metallurgy of the material being examined. If doubt still remains, or if the structure is too fine to observe, more sophisticated techniques must be implemented. With the equipment currently available, high-quality micrographs are easily produced. However, doing so requires careful attention to the preparation of a sample, its etching, and the use of the microscope. The sample must be properly prepared to ensure correct observation and interpretation of the microstructure without complications arising from artifacts due to inadequate preparation or etching of the sample.

An optical microscope can be used to examine polished or etched metallographic samples. Examination should be done before the sample is etched. Certain constituents are more readily observed in a polished sample because they are not obscured by etching detail. Inclusions, nitrides, certain carbides, and intermetallic phases can be readily observed without etching. Except for inclusions, the other phases may be more easily examined if some relief is introduced during final polishing. Samples that respond to polarized light, such as materials with noncubic crystal structures, are generally examined without etching. However, in most cases, etching must be performed to observe the microstructure. A general purpose etchant is normally used first to reveal the grain structure and the phases present, followed by selective etchants that attack or color specific phases of interest. Selective etchants are widely used for quantitative metallography, particularly if performed using an automated image analysis device. In either case, etching must be carefully performed to reveal the microstructure with clarity.

An annular combustor casing of a prototype engine made out of sheet metal was considered for the present study. This casing had 50 numbers of adaptor bosses mounted on its surface by welding. From the past experience, it has been observed that welded structures contain residual stresses due to thermal shrinkage of the weld metal and heat-affected zone (HAZ) in the welding cooling process. Therefore, necessary relief of stress needs to be carried out on welded structures. It has previously been observed that the failure of an aircraft is caused due to bursting of the combustion chamber [163,164]. The effect of the interaction between microstructural parameters and local stress concentration factors on component strength has been examined by researchers in greater detail through the concept of weld quality. It is observed that weld quality plays a major role in the strength of a fabricated component.

7.3 Details of the Test Component

During engine operation, the combustor casing is subjected to fatigue load. The fatigue load in service occurs due to start–stop operations and varying flight conditions. The fatigue loads may affect the combustor overpressure capability. The damage due to cyclic loads may manifest in the form of microstructure deterioration or initiation of cracks. In order to investigate the effect of damage induced in the casing by cyclic loads and its effect on the proof pressure capability, the casing was subjected to 2000 pressure cycles to evaluate its overpressure capability. During the experimental determination of the proof pressure of the casing, the pressure was increased till the casing eventually failed. A detailed investigation of the fracture surfaces was carried out using an SZM and an SEM to identify the mode of failure.

7.4 Experimental Procedure [165]

The experiment was conducted using a setup established to test both the cyclic load behavior and proof pressure of the component. Details of the test facility and the test setup were explained in Chapter 3.

The component was first subjected to cyclic pressure loading varying from 3 to 29.8 bar for 2000 cycles. At this stage, it was decided to increase the pressure further to evaluate the effect of 2000 cycles on the overpressure capability of this casing and the component pressure was continuously increased. Failure of the combustor chamber was observed at a pressure value of 50 bar. The burst pressure capability of a virgin combustor casing was demonstrated as 57 bar in Chapter 6. The burst pressure profile during the burst is shown in Figure 7.3. Subsequently, the casing was dismantled from the rig and subjected to detailed investigation to study the characteristics of the fracture.

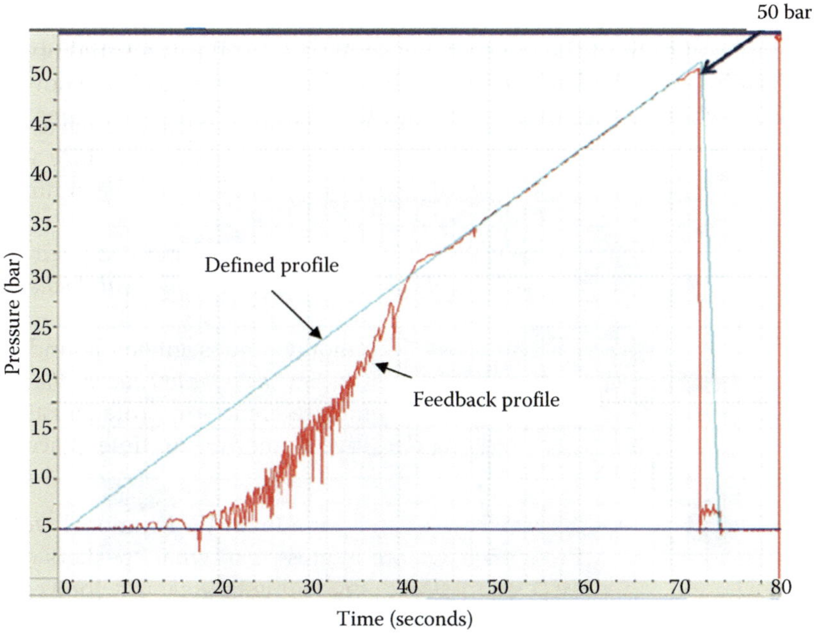

FIGURE 7.3
Test profile during pressure test. (With kind permission from Springer Science+Business Media: *J. Fail. Anal. Prev.*, Effect of fatigue on proof strength of gas turbine combustor casing, 13(1), 2013, 102–111, Sarangi, N., Panigrahi, S.K., Patel, S.K., Ramesh, P., Ganeshachar, M.D.)

7.5 Observation

7.5.1 Visual Observation

Visual observation revealed that two independent cracks were present in the failed diffuser casing. One axial crack was found to extend from the rear flange area to the upper flange area, covering the entire sheet metal casing, as shown in Figure 7.4. The length of this crack was found to be 36 cm. Another crack was found to have been initiated in the area of the rear flange. After initiation, this crack propagated circumferentially over a distance of about 48 cm in the rear flange area. It is to be noted here that the crack observed in the rear flange area overlapped with the previous crack that was observed after cyclic load testing.

7.5.2 Stereo Zoom Microscopy

The sectional view of the cracked region, to study the mode of failure, is shown in Figure 7.5a and b. An SZM with an auxiliary light source was used to highlight the fractured surface thoroughly to assess the extent of cracking, as well as to find the presence of other microcracks. The fractured surface macrographs of both of the cut sections showed a similar fractured appearance. Certain areas had a very smooth surface appearance, as shown in Figure 7.6a, whereas the surfaces around the region where the flanges were

FIGURE 7.4
Longitudinal and circumferential cracks in the combustor casing. (With kind permission from Springer Science+Business Media: *J. Fail. Anal. Prev.*, Effect of fatigue on proof strength of gas turbine combustor casing, 13(1), 2013, 102–111, Sarangi, N., Panigrahi, S.K., Patel, S.K., Ramesh, P., Ganeshachar, M.D.)

FIGURE 7.5
(a) Sectional view in axial direction and (b) sectional view in circumferential direction. (With kind permission from Springer Science+Business Media: *J. Fail. Anal. Prev.*, Effect of fatigue on proof strength of gas turbine combustor casing, 13(1), 2013, 102–111, Sarangi, N., Panigrahi, S.K., Patel, S.K., Ramesh, P., Ganeshachar, M.D.)

electron beam welded (EBW) with casings were relatively rough, as shown in Figure 7.6b. The lower cut section also showed presence of both smooth and rough regions, as shown in Figure 7.7.

7.5.3 Scanning Electron Microscopy

7.5.3.1 Fractography

The fracture surface as marked in Figure 7.8 of the axial crack region was observed under an SEM.

Three distinct fractography features can be seen:

1. Crystallographic faceting, indicating multiple fatigue crack origin (see Figure 7.9, at the outer wall)
2. Dimples coupled with finely spaced marks called "striations" (see Figure 7.10, middle of the wall thickness)
3. Dimple rupture region (see Figure 7.11, towards inner wall of the casing)

It was also observed from the fractographic features that the preferred crystallographic orientations of the fatigue striations were not affected

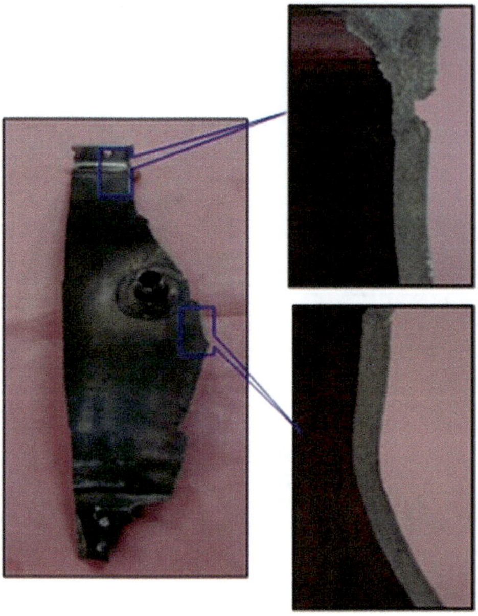

FIGURE 7.6
Appearance of fractured surfaces in the axial crack region. (With kind permission from Springer Science+Business Media: *J. Fail. Anal. Prev.*, Effect of fatigue on proof strength of gas turbine combustor casing, 13(1), 2013, 102–111, Sarangi, N., Panigrahi, S.K., Patel, S.K., Ramesh, P., Ganeshachar, M.D.)

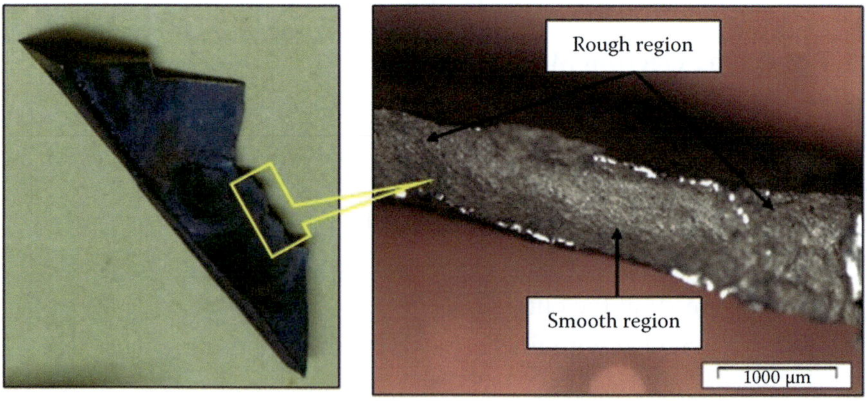

FIGURE 7.7
Appearance of fractured surfaces in the circumferential crack region. (With kind permission from Springer Science+Business Media: *J. Fail. Anal. Prev.*, Effect of fatigue on proof strength of gas turbine combustor casing, 13(1), 2013, 102–111, Sarangi, N., Panigrahi, S.K., Patel, S.K., Ramesh, P., Ganeshachar, M.D.)

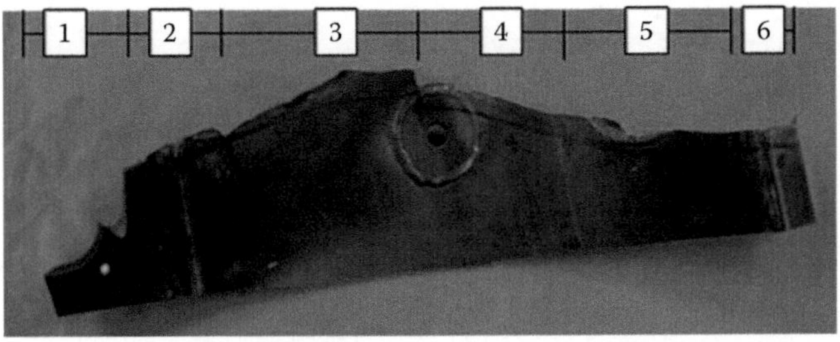

FIGURE 7.8
SEM fractograph of axial cracked region of combustor casing. (With kind permission from Springer Science+Business Media: *J. Fail. Anal. Prev.*, Effect of fatigue on proof strength of gas turbine combustor casing, 13(1), 2013, 102–111, Sarangi, N., Panigrahi, S.K., Patel, S.K., Ramesh, P., Ganeshachar, M.D.)

by the presence of dimples. In some sections, dimples were found to be distributed in patches. The fractured surface of the circumferential crack region, as observed in Figure 7.12, shows similar fractographic features as those observed in the axial crack region (see Figures 7.13 and 7.14).

7.5.3.2 Energy Dispersive X-Ray Analysis

Semiquantitative chemical analysis was carried out on the fractured surface using an energy dispersive x-ray (EDX) analyzer. The chemical composition of the fractured surface is listed in Table 7.1 and was found to match closely with the Inconel-718 material.

7.5.4 Optical Microscopy

The super alloy Inconel-718 used for this component is metallurgically complex and involves precipitation of several metastable phases. Further weldings of these alloys result in the generation of residual stress and variation in the hardness profile across the weld. A sample was cut from the axial fractured part, as shown in Figure 7.15, to verify the accuracy of postweld heat treatment (PWHT) given (720°C/8 hours/620°C/8 hours/Gas Fan Quench) to the combustor casing. Metallographic preparation, followed by etching the sample using concentrated $FeCl_3$, was carried out. Subsequently, the microstructure was observed under an optical microscope. Partial dissolution of dendrites along with laves phases was observed.

7.5.5 Microhardness Measurement

Since the weld section comprised two differently manufactured parts joined together, it was also necessary to see the hardness profile across the weld

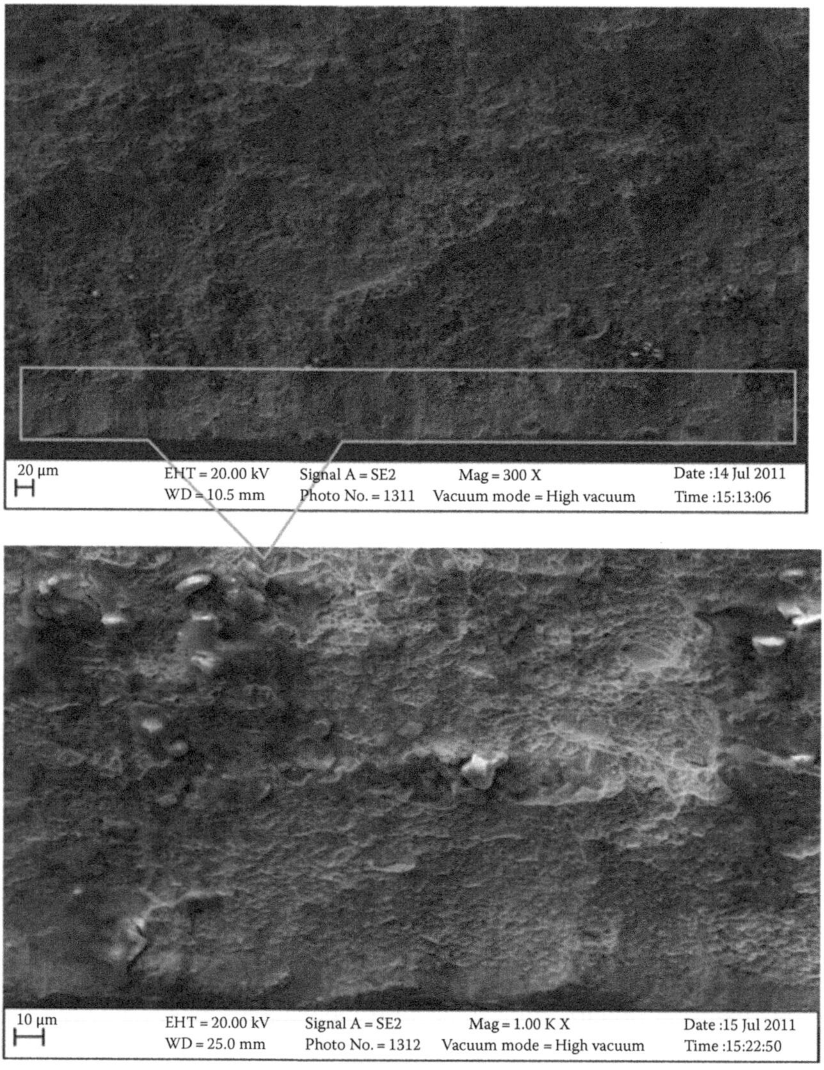

FIGURE 7.9
Crystallographic facets on the fractured surface of axial crack region. (With kind permission from Springer Science+Business Media: *J. Fail. Anal. Prev.*, Effect of fatigue on proof strength of gas turbine combustor casing, 13(1), 2013, 102–111, Sarangi, N., Panigrahi, S.K., Patel, S.K., Ramesh, P., Ganeshachar, M.D.)

zone. Hence, a microhardness measurement under a 300 g load was carried out on the sample used for optical microscopy. The hardness profile obtained, as shown in Figure 7.16, varies around 42–48 RC (Rockwell hardness number in scale-C). Hence, no significant hardness variation was observed between the forged area and the ring rolled area.

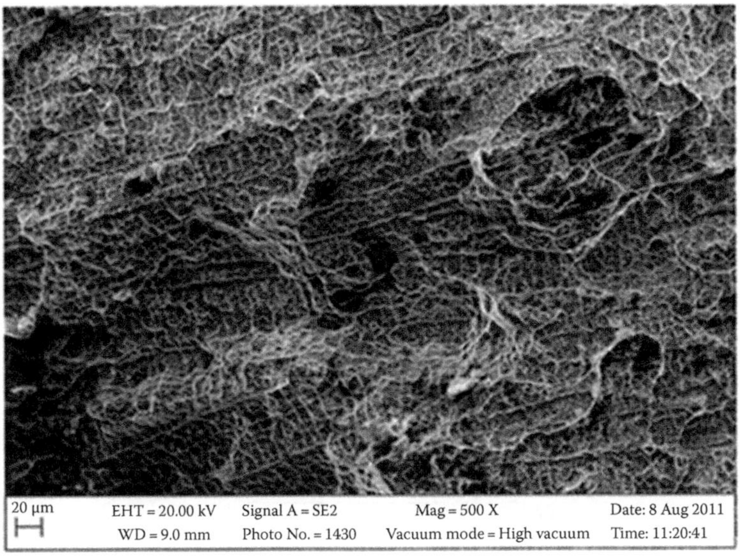

FIGURE 7.10
Microstructures showing the coupling of dimples with striations in the axial crack region. (With kind permission from Springer Science+Business Media: *J. Fail. Anal. Prev.*, Effect of fatigue on proof strength of gas turbine combustor casing, 13(1), 2013, 102–111, Sarangi, N., Panigrahi, S.K., Patel, S.K., Ramesh, P., Ganeshachar, M.D.)

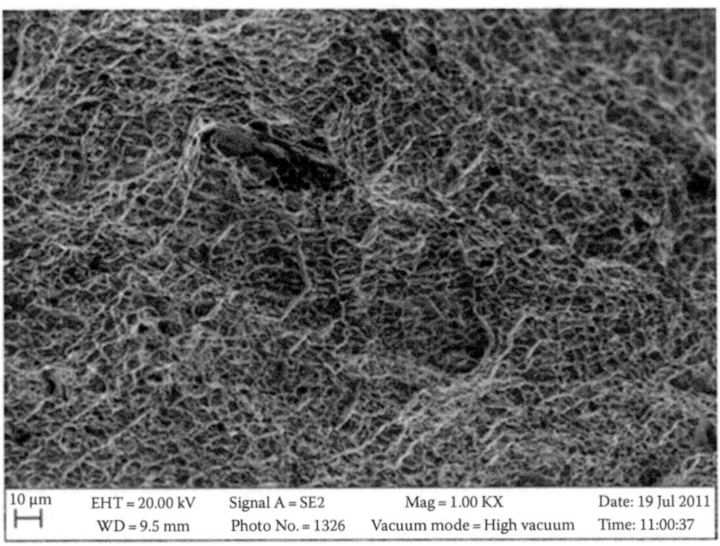

FIGURE 7.11
Microstructures showing the presence of dimples in the axial crack region. (With kind permission from Springer Science+Business Media: *J. Fail. Anal. Prev.*, Effect of fatigue on proof strength of gas turbine combustor casing, 13(1), 2013, 102–111, Sarangi, N., Panigrahi, S.K., Patel, S.K., Ramesh, P., Ganeshachar, M.D.)

FIGURE 7.12
SEM fractograph of circumferential crack region. (With kind permission from Springer Science+Business Media: *J. Fail. Anal. Prev.*, Effect of fatigue on proof strength of gas turbine combustor casing, 13(1), 2013, 102–111, Sarangi, N., Panigrahi, S.K., Patel, S.K., Ramesh, P., Ganeshachar, M.D.)

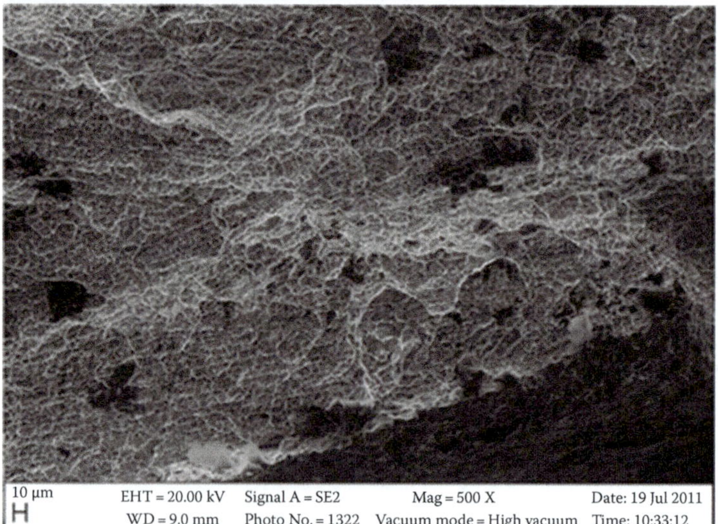

FIGURE 7.13
Microstructure showing the coupling of dimples with striations in the circumferential crack region. (With kind permission from Springer Science+Business Media: *J. Fail. Anal. Prev.*, Effect of fatigue on proof strength of gas turbine combustor casing, 13(1), 2013, 102–111, Sarangi, N., Panigrahi, S.K., Patel, S.K., Ramesh, P., Ganeshachar, M.D.)

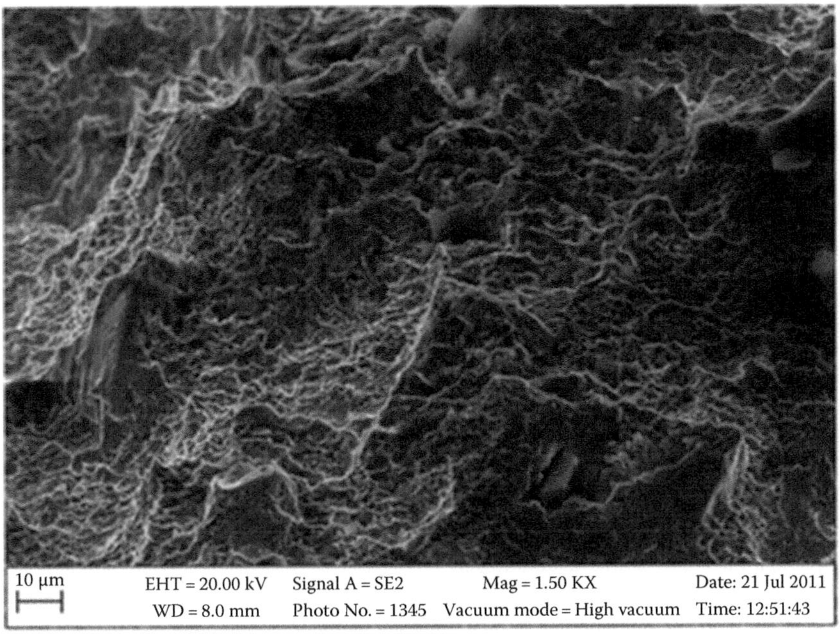

FIGURE 7.14
Microstructures showing the presence of dimples on the fractured surface of circumferential crack region. (With kind permission from Springer Science+Business Media: *J. Fail. Anal. Prev.*, Effect of fatigue on proof strength of gas turbine combustor casing, 13(1), 2013, 102–111, Sarangi, N., Panigrahi, S.K., Patel, S.K., Ramesh, P., Ganeshachar, M.D.)

TABLE 7.1
Chemical Composition of the Fractured Surface

	Weight by %	
Elements	**Present**	**Inconel-718 [8]**
Aluminum	0.9	0.3–0.7
Titanium	1.43	0.75–1.15
Chromium	18.57	17–21
Iron	19.24	14–24
Nickel	Balance	Balance
Niobium	5.04	4.75–5.50
Molybdenum	3.11	2.8–3.3

Source: With kind permission from Springer Science+Business Media: *J. Fail. Anal. Prev.*, Effect of fatigue on proof strength of gas turbine combustor casing, 13(1), 2013, 102–111, Sarangi, N., Panigrahi, S.K., Patel, S.K., Ramesh, P., Ganeshachar, M.D.

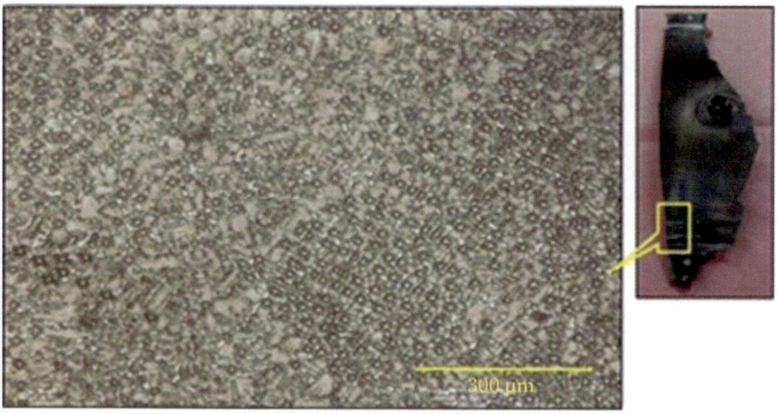

FIGURE 7.15
Micrograph showing the dissolution of dendrites and presence of laves phases. (With kind permission from Springer Science+Business Media: *J. Fail. Anal. Prev.*, Effect of fatigue on proof strength of gas turbine combustor casing, 13(1), 2013, 102–111, Sarangi, N., Panigrahi, S.K., Patel, S.K., Ramesh, P., Ganeshachar, M.D.)

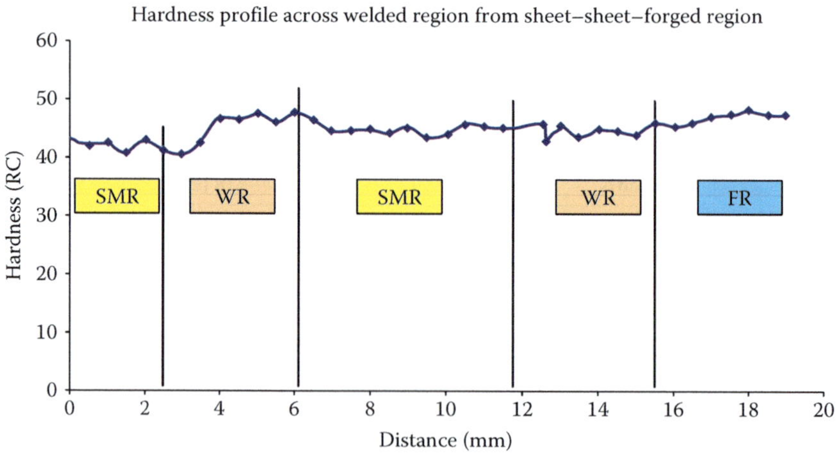

FIGURE 7.16
Microhardness profile across the welded area: SMR, sheet metal region; WR, weld region; FR, forged region. (With kind permission from Springer Science+Business Media: *J. Fail. Anal. Prev.*, Effect of fatigue on proof strength of gas turbine combustor casing, 13(1), 2013, 102–111, Sarangi, N., Panigrahi, S.K., Patel, S.K., Ramesh, P., Ganeshachar, M.D.)

7.6 Discussions

During fabrication of the component, the formed sheet metal part was welded at both ends to the front flange and the rear flange. The welding of the front flange area to the casing was followed by its welding to the rear flange using EBW, which generated a concentration of residual stress. To relieve the residual stress, the component was subjected to PWHT (720 C/8hrs/620/8hrs/Gas Fan Quench).

In the present case, by observing the contour of the crack path it was confirmed that the EBW process to join the three parts (upper flange, case, and lower flange) was found to be intact and had not directly contributed to the failure. This shows that the region joined by EBW had sufficient strength to withstand the proof pressure.

The tungsten inert gas (TIG) welding of bosses in different areas resulted in microstructure heterogeneity at these locations. Further, the area of the axial crack contained a section of the TIG-welded area. The fractured surface of both areas showed the presence of a very smooth region under SZM observation, which indicates that failure occurred in a progressive mode. The presence of crystallographic facets at the outer circumference of the casing, as observed under SEM, clearly indicates multiple points of crack origination. The presence of striations coupled with dimples further confirmed the contribution due to fatigue failure. However, the final failure occurred due to overload, as indicated by the presence of dimples. The progressive fatigue damage may have affected the overall loading capability of the component. Similar fractographic features were observed in both the crack areas. It is to be noted here that both cracks were initiated during cyclic testing only. Semiquantitative chemical analysis confirmed the absence of any extraneous elements at the fracture locations. Optical microscopy indicated the partial dissolution of dendrites, along with the presence of deleterious laves phases, which infers the inability of PWHT completely to dissolve dendrites and deleterious laves phases. However, it is to be noted here that, after PWHT, the accepted hardness level of around 42 RC was acquired, as it is evident from the hardness profile obtained across the weld region, that is, 42–48 RC. The PWHT provided could not completely dissolve the dendrites or eliminate the laves phases. Once the fatigue crack had been initiated, the presence of laves phases promoted the propagation of the fatigue cracks.

7.7 Summary

A fatigue loaded diffuser casing was under investigation to evaluate its overloading capability. This was undertaken by subjecting it to a proof pressure test where it was observed to fail at 50 bar due to the development of two cracks: one axial and the other circumferential. The presence of very smooth regions indicates that the fatigue failure was progressive. The fracture surfaces of both regions displayed similar fractographic features. This indicated that cracks began to form in both areas during cyclic pressure testing. The loading profile selected for the casing pressure test, which closely represented the operational loads, clearly showed that fatigue loading affects the overpressure capability of the casing.

8

Conclusions

8.1 Certification and Acceptance of the Combustor Casing

An aero engine structural component must be certified by the appropriate regulating body to ensure its compliance with airworthiness requirements before the component is fitted in an aircraft. The certificate reflects a determination made by a regulating body that the component is manufactured according to an approved design and complies with the safety requirements of the component, as well as its life requirement. This book provides a comprehensive methodology for assessing the durability and robustness of an aero engine combustor casing through an integrated approach involving experimental and numerical means of analysis. The combustor casing of an aero engine is one of the most critical parts of the engine. It is formed from thin metal sheets and has a doubly curved surface, which is curved about z-axis as well as y-axis. The mounting of bosses to locate igniters, atomizers, and instrumentation on the surface of the combustor make the component of highly critical importance. The bosses, which are welded to the combustor surface, raise the stresses the combustor surface undergoes and affect the life and strength of the casing. It is a mandatory requirement for certification to be given that the life and strength of the combustor casing be demonstrated by means of a full-scale rig test prior to its fitting in the engine. The modified combustor casing demonstrated the necessary fatigue life as well as the proof capability required for its airworthiness requirement. The component was certified, has been accepted by the certifying authorities, is fitted in one of India's indigenously developed gas turbine engines, and is working satisfactorily.

8.2 Specific Conclusions

A state-of-the-art servo hydraulic pressure test facility working on a closed loop system was developed. A test setup was designed and fabricated to adapt the combustor casing for the test facility. Hydraulic oil was used as the pressurizing media and the tests were planned at room temperature; the cyclic pressure test was conducted on a full-scale combustor casing made out of sheet metal to evaluate its fatigue life. The component failed prematurely at 1825 cycles as opposed to the desired 7000 cycles.

The critical region in the casing was identified and a detailed analysis was carried out to predict the life of the casing. Various factors which affect the fatigue life at the boss-shell weld in the casing were identified and considered for improved correlation of the estimation of the casing's life. These factors were refined based on the casing fatigue life obtained from the testing.

Based on the study, the manufacturing processes were improved, as was the configuration of the bosses. It was recommended that the combustor casing is constructed by machining forged billets to the required thickness.

The manufacturing methodology was changed based on this recommendation and a new combustor casing was accordingly manufactured. The new modified combustor casing could demonstrate 28,000 pressure cycles compared with the required 7,000 cycles. The combustor remained in a very good condition without any deformation or any crack, even after 28,000 cycles. Therefore, the study gave invaluable input for the improvement of a highly critical component in an aero engine.

One separate new modified combustor casing was subjected to the proof pressure test to demonstrate its safety and overload capability in the same pressure test facility. This time, the combustor was subjected to 57 bar, which is double the maximum operating pressure, for a duration of 300 seconds. The component was inspected before and after the test and no distortion was observed in the component. The new combustor casing demonstrated the required safety and overload capabilities.

The test also provided invaluable information that validated the new manufacturing methodology; namely, machining the component from forged billets.

An attempt was made to find the effect of fatigue on the overload capability. In this study, the combustor was subjected to 2000 pressure cycles and then subjected to overload. The pressure was increased till the component ruptured. It was observed that the combustor failed at 50 bar, which is 7 bar less than its required capability. The failed region was investigated and it was confirmed that the failure has occurred due to fatigue followed by overload. This study demonstrated that fatigue load affects the overload capability of the component.

8.3 Scope for Further Work

The last five decades have witnessed revolutionary efforts in the development of new methodologies and design concepts resulting in enhanced and multirole capable aero engines. The axial gas turbine engine was a major improvement over the centrifugal compressor turbine engine, such as the von Ohain or the Whittle engines. All large jet engines today are axial gas turbine engines.

In a gas turbine engine, the combustor or combustion chamber is fed high pressure air by the compression system. The combustor then heats this air at constant pressure. After heating, air passes from the combustor through the nozzle guide vanes to the turbine and then to the nozzle for the required thrust. The combustor must contain and maintain stable combustion despite very high air flow rates. To do so, the combustors are carefully designed, first, to mix and ignite the air–fuel mixture and, then, to mix in more air to complete the combustion process. In present day aero engines, afterburners are often considered another type of combustor. The combustors play a crucial role in determining many of an engine's operating characteristics, such as fuel efficiency, levels of emissions, and transient response (the response to changing conditions, such as fuel flow and air speed). During operation, it is subjected to maximum engine pressure and relatively high temperature. Military standards make it mandatory to demonstrate the life and strength of a combustor by testing a prototype of the combustor. Although there has been a significant increase in the use of advanced software and technologies for the validation of component design, the stringent requirements laid down by the airworthiness authorities specify that a sufficient number of components need to be tested in the exact engine environment for a designer to have full confidence in the component's design.

Therefore, research can be carried further in those emerging fields. In view of the considerations noted, the following areas of work related to the present investigation are suggested for further study:

- Development of an empirical model for the assessment of combustor life
- Metallographic studies, including crack initiation and crack growth

References

1. Andrew, N. (2004) *Frank Whittle: Invention of the Jet*, Icon Books Ltd., ISBN: 1-84046-538-7.
2. Whittle, F. (1953) *Jet: The Story of a Pioneer*, Frederick Muller Ltd.
3. Whittle, F. (1981) *Gas Turbine Aero-Thermodynamics: With Special Reference to Aircraft Propulsion*, Pergamon, ISBN: 978-0-08-026718-0, pp. 8–25.
4. Warsitz, L. (April 2008) *The First Jet Pilot*, Pen and Sword Books, South Yorkshire, U.K.
5. Buckingham, E. (1902) Theory of thermodynamics, *Bulletin of the American Mathematic Society*, 9: 173–175.
6. Thermonet (n.d.), Department of Mechanical Engineering, University of Texas, Austin, TX, http://www.wiley.com/college/moran/CL_0471465704_S/user/tutorials/tutorial9/tut9h_content.html.
7. Shenyang Blower Works Group (n.d.), Katy, TX, http://www.sbw-turbo.com/ewebeditor/UploadFile/20151214145516634.jpg.
8. Concepts NREC (n.d.), White River Junction, VT, http://www.conceptsnrec.com/getmedia/e3c0a482-ed58-40f6-8955-d8cc648d13a2/gallery-comp11.aspx.
9. Sleipnir, Lycoming T 53 gas turbine, October 2005, https://en.wikipedia.org/wiki/File:Turboprop_T-53.jpg, last modified on March 6, 2017.
10. Lefebvre, A.H., Ballal, D.R. (2010) *Gas Turbine Combustion: Alternative Fuels and Emissions*, 3rd edn., CRC Press, New York, pp. 9–14.
11. Ganesan, V. (2006) *Gas Turbines*, 2nd edn., Tata McGraw-Hill Publishing Company Ltd., New Delhi, India, pp. 211–282.
12. General Electric, Turbojet, OH (n.d.), https://en.wikipedia.org/wiki/General_Electric_J85, last modified on November 27, 2016.
13. Stack Overflow Talent Orders (n.d.), New York, NY, https://www.reddit.com/r/MachinePorn/comments/320jj1/a_nozzle_on_a_mcdonnell_douglas_f15_eagle_at_the evergreen aviation and space museum in McMinville/.
14. Stack Overflow (n.d.), New York, NY, http://space.stackexchange.com/questions/1171/efficient-types-of-nozzles-used-in-rockets.
15. Shakariyants, S.A., Van Buijtenen, J.P., Visser, W.P.J. (2004) Generic geometry definition of the aircraft engine combustion chamber, GT2004-53522, *Proceedings of ASME Turbo Expo 2004*, Vienna, Austria.
16. Nima, P., Michael, P. (2006) Development of a combustion chamber design methodology and automation of the design process, *25th International Congress of the Aeronautical Sciences, Hamburg, Germany*, pp. 1–10.
17. Khandelwal, B., Karakurt, A., Sekaran, P.R., Sethi, V., Singh, R. (July 2013) Hydrogen powered aircraft: The future of air transport, *Progress in Aerospace Sciences*, 60: 45–59.
18. Bin, H., Yong, H., Fang, W., Xie, F. (January 2013) CFD predictions of LBO limits for aero-engine combustors using fuel iterative approximation, *Chinese Journal of Aeronautics*, 26: 74–84.

19. Muller, U., Hoschler, K., Gerendas, M., Bauer, H.J., Schoth, U. (2002) Mechanical analysis of an aero-engine combustor under operation conditions using a unified constitutive material model for deformation simulation, *European Structural Integrity Society*, 29: 351–360.
20. Jin, Y., Xiaomin, H.E., Jiang, B., Zejun, W.U., Guoyu, D. (2012) Design and performance of an improved trapped vortex combustor, *Chinese Journal of Aeronautics*, 25(6): 864–870.
21. Nima, P., Michael, P. (2006) Development of a combustion chamber design methodology and automation of the design process, *25th International Congress of the Aeronautical Sciences*, Hamburg, Germany, pp. 1–10.
22. Rolls-Royce Plc. (1996) *The Jet Engine*, 5th edn., Renault Printing Co., Birmingham, Great Britain, pp. 35–44.
23. Das, S., Babu, S.N.N., Ghosh, S.N. (2004) Advances in materials for Kaveri engine development programme, *Seventh National Conference on Air Breathing Engines and Aerospace Propulsion*, NCABE-2004, IIT, Kanpur, India, pp. 399–411.
24. MIL-E-5007D. (October 15, 1973) Military specifications, General specifications for engines, aircraft, turbojet and turbofan.
25. INCONEL ALLOY 718. (1985) *Material Broucher*, 4th edn., INCO Alloys International, pp. 1–28.
26. Thomas, A., El-Wahabi, M., Cabrera, J.M., Prado, J.M. (2006) High temperature deformation of Inconel 718, *Journal of Materials Processing Technology*, 177(1–3): 469–471.
27. Huda, Z., Edi, P. (2013) Materials selection in design of structures and engines of supersonic aircrafts: A review, *Materials & Design*, 46: 552–560.
28. Hong, J.K., Park, J.H., Park, N.K., Eom, I.S., Kim, M.B., Kang, C.Y. (2008) Microstructures and mechanical properties of Inconel 718 welds by CO_2 laser welding, *Journal of Materials Processing Technology*, 201(1–3): 515–520.
29. Damodaram, R., Sundara Raman, S.G., Rao, K.P. (2013) Microstructure and mechanical properties of friction welded alloy 718, *Materials Science and Engineering*, 560: 781–786.
30. Koff, B.L. (2004) Gas turbine technology evolution: A designer's perspective, *Journal of Propulsion and Power*, 20(4): 577–589.
31. Koff, B.L. (1884) Designing for durability in fighter engines, *International Journal of Turbo and Jet Engines*, 1: 209–222.
32. Rudrapatna, N.S., Peterson, B.H., Daniel, G. (2011) An experimental system for assessing combustor durability, *Journal of Engineering for Gas Turbines and Power*, 133: 042103-1–042103-6.
33. Taylor, D., Hoey, D. (2009) High cycle fatigue of welded joints, *International Journal of Fatigue*, 31: 20–27.
34. Toyosada, M., Gotoh, K., Niwa, T. (2004) Fatigue life assessment for welded structures without initial defects: An algorithm for predicting fatigue crack growth from a sound site, *International Journal of Fatigue*, 26: 993–1002.
35. Jen, Y.M., Chang, L.Y., Fang, C.F. (2008) Assessing the fatigue life of butt-welded joints under oblique loading by using local approaches, *International Journal of Fatigue*, 30: 603–613.
36. Dogan, B. (2010) Fitness of service of welded components under creep and creep-fatigue loading, *International Journal of Pressure Vessels and Piping*, 87: 656–663.

37. Barsoum, Z.G. (2009) Fatigue of high strength steel joints welded with low temperature transformation consumables, *Journal of Engineering Failure Analysis*, 16: 2186–2194.
38. Kainuma, S., Mori, T. (2008) A study on fatigue crack initiation point of load-carrying fillet welded cruciform joints, *International Journal of Fatigue*, 30: 1669–1677.
39. Pakandam, F., Farahani, V.A. (2011) Fatigue damage assessment of various welded joints under uniaxial loading based on energy method, *International Journal of Fatigue*, 33: 519–528.
40. Chapetti, M.D., Jaureguizahar, L.F. (2012) Fatigue behavior prediction of welded joints by using an integrated fracture mechanics approach, *International Journal of Fatigue*, 43: 43–53.
41. Sehgal, R., Angra, S., Sharma, V. (2007) Failure cause analysis of welded joints, *Indian Journal of Engineering & Materials Sciences*, 14: 24–30.
42. Zhang, Y.H., Stephen, J.M. (2009) Fatigue life prediction for toe ground welded joints, *International Journal of Fatigue*, 31: 1124–1136.
43. Tao, C.H., Zhang, D., Xi, N.S. (2000) Explosion of a combustion chamber shell, *Journal of Engineering Failure Analysis*, 7: 1–10.
44. Lin, Y.C., Chen, X.M., Liu, Z.H., Chen, Z. (November 2012) Investigation of uniaxial low-cycle fatigue failure behavior of hot rolled AZ91 magnesium alloy, *International Journal of Fatigue*, 48: 122–132.
45. Evans, W.J., Screech, J.E., Williams, S.J. (2008) Thermo-mechanical fatigue and fracture of Inco718, *International Journal of Fatigue*, 30: 257–267.
46. Evans, W.J., Jones, J.P., Bache, M.R. (2003) High temperature fatigue/creep environment interactions in compressor alloys, *Journal of Engineering and Gas Turbine Power*, 125: 246–251.
47. Makkonen, M. (2009) Predicting the total fatigue life in metals, *International Journal of Fatigue*, 31: 1163–1175.
48. Roy, S.C., Goyal, S., Sandhya, R., Ray, S.K. (2012) Low cycle fatigue life prediction of 316 L(N) stainless steel based on cyclic elasto-plastic response, *Nuclear Engineering and Design*, 253: 219–225.
49. Zaletelj, H., Fajdiga, G., Nagode, M. (2013) Low cycle fatigue of welded joints, *Materials Science and Engineering*, 564(1): 478–484.
50. Yung, L.L., Jwo, P., Richard, B.H., Mark, E.B. (2005) *Fatigue Testing and Analysis*, Elsevier Butterworth-Heinemann, pp. 57–102.
51. Zhang, Y., Hu, C.L., Zhao, Z., Li, A.P., Xu, X.L., Shi, W.B. (2013) Low cycle fatigue behavior a Cr–Mo–V matrix type high speed steel used for cold forging, *Materials & Design*, 44: 612–621.
52. Yung, L.-L., Jwo, P., Richard, H., Mark, E.B. (2005) *Fatigue Testing and Analysis: Theory and Practice*, Elsevier Butterworth, Oxford, U.K., pp. 57–76.
53. Goswamy, T. (1997) Low cycle fatigue life prediction, *International Journal of Fatigue*, 19: 109–115.
54. Mamiya, E.N., Castro, F.C., Algarte, R.D., Araujo, J.A. (2011) Multiaxial fatigue life estimation on a piecewise ruled S–N surface, *International Journal of Fatigue*, 33: 529–540.
55. James, M.N., Paterson, A.E. (1997) Fatigue performance of 6261-T6 aluminum alloy—Constant and variable amplitude of loading of parent plate and welded specimens, *International Journal of Fatigue*, 19: 109–118.

56. Corran, R.S.J., Williams, S.J. (2007) Lifing methods and safety criteria in aero gas turbines, *Engineering Failure Analysis*, 14: 518–528.
57. Hicks, M.A., Pickard, A.C. (1988) Life prediction in turbine engines and the role of small cracks, *Materials Science and Engineering*, 103(1): 43–48.
58. Schutz, W. (1996) A history of fatigue, *Engineering Fracture Mechanics*, 54(2): 263–300.
59. Cruse, T.A., Meyer, T.G. (1979) A cumulative fatigue damage model for gas turbine engine disks subjected to complex mission loading, *Journal of Engineering for Power*, 101: 563–571.
60. Biswas, S., Ganeshachar, M.D., Sridhar, B.R., Ramachandra, S. (2010) Failure analysis of an air turbine starter, *Journal of Failure Analysis and Prevention*, 10: 259–263.
61. Bhaumik, S.K., Sujata, M., Venkataswamy, M.A., Parameswara, M.A. (2006) Failure of a low pressure turbine rotor blade of an aero engine, *Journal of Engineering Failure Analysis*, 13: 1202–1219.
62. Sunder, R., Seetharam, S.A., Bhaskaran, T.A. (1984) Cycle counting for fatigue crack growth analysis, *International Journal of Fatigue*, 3: 147–156.
63. Sunder, R., Dash, P.K. (1982) Measurement of fatigue crack closure through electron microscopy, *International Journal of Fatigue*, 4(2): 97–105.
64. Nelson, D.V., Fuchs, H.O. (1977) *Predictions of Cumulative Fatigue Damage Using Condensed Load Histories under Complex Loading*, Society of Automotive Engineering Inc., Warrendale, PA, p. 163.
65. Ozawa, Y., Tochihara, Y., Mori, N., Yuri, I., Sato, J., Kagawa, K. (2003) Test results of a catalytically assisted combustor for a gas turbine, *Catalysis Today*, 83(1–4): 247–255.
66. Chenyu Wei, C., Shusheng Zang, S. (2013) Experimental investigation on the off-design performance of a small-sized humid air turbine cycle, *Applied Thermal Engineering*, 51(1–2): 166–176.
67. Savas, T.P. (2000) Fatigue and stress analysis of a novel test coupon geometry developed for hydraulic pressure impulse testing, *Journal of Testing and Evaluation*, 28(5): 8.
68. Wang, H., Wang, Z., Sun, M. (2013) Experimental study of oscillations in a scramjet combustor with cavity flame holders, *Experimental Thermal and Fluid Science*, 45: 259–263.
69. Singh, I.V., Mishra, B.K., Bhattacharya, S., Patil, R.U. (2012) The numerical simulation of fatigue crack growth using extended finite element method, *International Journal of Fatigue*, 36: 109–119.
70. Shinohara, K., Qiang, Y. (2011) Fatigue life evaluation accuracy of power devices using finite element method, *International Journal of Fatigue*, 33: 1221–1234.
71. Collins, J.A. (1993) *Failure of Materials in Mechanical Design*, John Wiley & Sons, New York, pp. 178–332.
72. Asquith, G., Pickard, A.C. (1998) Fatigue testing of gas turbine components, *International Journal of High Temperature Technology*, 6: 131–143.
73. Huang, W., Zhen-guo Wang, Z.G., Yan, L., Liu, W. (2012) Numerical validation and parametric investigation on the cold flow field of a typical cavity-based scramjet combustor, *Acta Astronautica*, 80: 132–140.
74. Khan, I.A., Ahuja, P., Satpute, S., Khan, M.A., Bhasin, V., Vaze, K.K., Ghosh, A.K. et al. (2011) Fracture investigations on piping system having large through-wall circumferential crack, *International Journal of Pressure Vessels and Piping*, 88: 223–230.

75. Jacobsson, L., Persson, C., Melin, S. (2009) Thermo-mechanical fatigue crack propagation experiments in Inconel 718, *International Journal of Fatigue*, 31: 1318–1326.
76. Filippini, M., Foletti, S., Pasqueo, G. (2010) Assessment of multiaxial fatigue life prediction methodologies for Inconel 718, *Procedia Engineering*, 2: 2347–2356.
77. Gustafsson, D., Moverare, J., Simonsson, K., Johansson, S., Hornqvist, M., Mansson, T., Sjostrom, S. (2011) Fatigue crack growth behaviour of Inconel 718—The concept of a damaged zone caused by high temperature hold times, *Procedia Engineering*, 10: 2821–2826.
78. Chen, Q., Kawagoishi, N., Nisitani, H. (2000) Evaluation of fatigue crack growth rate and life prediction of Inconel 718 at room and elevated temperatures, *Materials Science and Engineering*, A277: 250–257.
79. Witek, L. (2006) Failure analysis of turbine disc of an aero engine, *Engineering Failure Analysis*, 13: 9–17.
80. Kim, H. (2009) Study of the fracture of the last stage blade in an aircraft gas turbine, *Engineering Failure Analysis*, 16: 2318–2324.
81. Epaarachchi, J.A. (2005) A new cumulative fatigue damage model for glass fibre reinforced plastic composites under step/discrete loading, *Journal of Applied Science and Manufacturing*, 36(9): 1236–1245.
82. Todinov, M.T. (2001) Necessary and sufficient condition for additivity in the sense of the Palmgren–Miner rule, *Journal of Computational Materials Science*, 21(1): 101–110.
83. Shimokawa, T., Tanaka, S. (1980) A statistical consideration of Miner's rule, *International Journal of Fatigue*, 2(4): 165–170.
84. Kauzlarich, J.J. (1989) The Palmgren–Miner rule derived, *Tribology Series*, 14: 175–179.
85. Sun, Y.S. (1994) Revised Miner's rule and its application in calculating equivalent loads for components, *Reliability Engineering & System Safety*, 43(3): 319–324.
86. Nian, Z.C., Wang, G., Soares, C.G. (2011) Palmgren-Miner's rule and fracture mechanics based inspection planning, *Journal of Engineering Fracture Mechanics*, 78: 3166–3182.
87. Yao, W.X. (2007) A cumulative fatigue damage rule under the alternative of corrosion or cyclic loading, *Acta Metallurgica Sirica (English Letters)*, 20(1): 65–71.
88. Hosoi, A., Kawada, H., Yoshino, H. (2006) Fatigue characteristics of quasi-isotropic CFRP laminates subjected to variable amplitude cyclic two-stage loading, *International Journal of Fatigue*, 28(10): 1284–1289.
89. Kang, H.T., Lee, Y.L. (2012) Fatigue life prediction methods of seam-welded joints, In: *Metal Fatigue Analysis Handbook—Practical Problem-Solving Techniques for Computer-Aided Engineering*, Elsevier, the Netherlands, pp. 383–427.
90. Benabou, L., Sun, Z., Dahoo, P.R. (2013) A thermo-mechanical cohesive zone model for solder joint lifetime prediction, *International Journal of Fatigue*, 49: 18–30.
91. Conle, F.A., Bonnen, J.J.F. (2008) Using the Neuber plot to account for the effects of scatter, corrosion andwelding in strain-life fatigue test data, *Proceedings of the Eighteenth International Offshore and Polar Engineering Conference*, Vancouver, British Columbia, Canada, July 6–11, 2008.
92. Schijve, J. (2012) Fatigue predictions of welded joints and the effective notch stress concept, *International Journal of Fatigue*, 45: 31–38.

93. Mazumdar, P.K., Conrad, H. (1985) A strain criterion for fatigue threshold based on Neuber's approximation, *Engineering Fracture Mechanics*, 21(3): 573–577.
94. Cristea, M.E., Beretta, S., Altamura, A. (2012) Fatigue limit assessment on seamless tubes in presence of inhomogeneities: Small crack model vs. full scale testing experiments, *International Journal of Fatigue*, 41: 150–157.
95. Terrel, J.B. (1989) Use of Neuber's rule to estimate the fatigue life of notched specimens of ASME SA 106-B steel piping in 288°C air, *International Journal of Pressure Vessels and Piping*, 40(1): 17–40.
96. Karakas, O. (2013) Consideration of mean-stress effects on fatigue life of welded magnesium joints by the application of the Smith–Watson–Topper and reference radius concepts, *International Journal of Fatigue*, 49: 1–17.
97. Coffin, L.E. (1954) A study of the effects of cyclic thermal stresses on a ductile metal, *Transactions of the ASME*, 76: 931–950.
98. Manson, S.A. (1954) Technical Note 2933, National Advisory Committee for Aeronautics, pp. 1–34.
99. Benasciutti, D. (2012) On thermal stress and fatigue life evaluation in work rolls of hot rolling mill, *Journal of Strain Analysis*, 47: 297–312.
100. Radhakrishnan, V.M. (1992) On the bilinearity of the Coffin–Manson low cycle fatigue relationship, *International Journal of Fatigue*, 14(5): 305–311.
101. Nieslony, A., Doki, C., Kaufmann, H., Krug, P. (2008) New method for evaluation of the Manson–Coffin–Basquin and Ramberg–Osgood equations with respect to compatibility, *International Journal of Fatigue*, 30(10): 1967–1977.
102. Jankovic, M. (2001) The application of some approximate solutions of stress and strain concentration for life estimation in the low cycle fatigue region, *The Scientific Journal Facta Universitatis*, 8: 1025–1032.
103. Willcox, M., Downes, G. (2003) *A Brief Description of NDT Techniques*, Insight NDT Equipments Ltd., pp. 1–20.
104. Mechanical Testing (n.d.), AE3145 Test Machines, www.soliton.ae.gatech.edu/people/jcraig/classes/ae3145/Lab3/test-machines.pdf.
105. Hooper, M.N., Hutchinson, A.R., Broughton, J.G., Taylor, M.A. (2012) Development of a novel test rig for the evaluation of aircraft fuel tank sealant, *Journal of Testing and Evaluation*, 40(1): 177–183.
106. Nonaka, I., Ito, T., Takemasa, F., Saitou, K., Miyachi, Y., Fujita, A. (2007) full size internal pressure creep test for welded P91 hot reheat elbow, *International Journal of Pressure Vessels and Piping*, 84(1–2): 97–103.
107. Hooreweder, V.B., Moens, D., Boonen, R., Sas, P. (2012) Design and simulation of a novel multiaxial fatigue test rig, *International Journal of Experimental Mechanics*, 52(5): 513–524.
108. Yun, G.J., Abdullah, A.B.M., Binienda, W. (2012) Development of a closed loop high cycle resonant fatigue testing system, *Journal of Experimental Mechanics*, 52: 275–288.
109. Pach, E., Korin, I., Ipina, J.P. (2012) Simple fatigue testing machine for fibre-reinforced polymer composite, *Journal of Experimental Techniques*, 36: 76–82.
110. Bush, R.W. (1985) Effect of dither on strain-controlled fatigue and tensile results obtained with a closed-loop servo hydraulic testing machine, *Journal of Testing and Evaluation*, 13(6): 13–16.
111. Moog Inc. (n.d.), New York, NY, http://infokitabersama123.blogspot.in/2016/03/mengenal-prinsip-kerja-moog-servo-valve.html.

112. Sarangi, N., Panigrahi, S.K., Chandrasekhar, U. (2014) Experimental life evaluation of an annular combustor casing of an aero gas turbine engine, *Journal of Testing and Evaluation*, 42(3): 629–638.
113. Music, O., Allwood, J.M., Kawai, K. (2010) A review of mechanics of metal spinning, *Journal of Materials Processing Technology*, 210(1): 3–23.
114. Palten, H., Palten, D. (2002) Metal spinning: From Ancient Art to High-Tech Industry, http://www.leifeldspinning.com/pdf_videos_news_releases/metalforming_mag9_02.pdf. Accessed on September 2002.
115. Degarmo, E.P., Black, J.T., Kohser, R.A. (2003) *Materials and Processes in Manufacturing*, 9th edn., Wiley, pp. 140–172.
116. Wang, L., Long, H. (2011) Investigation of material deformation in multi-pass conventional metal spinning, *Materials & Design*, 32(5): 2891–2899.
117. Quigley, E., Monaghan, J. (2002) Enhanced finite element models of metal spinning, *Journal of Materials Processing Technology*, 121(1): 43–49.
118. Low Carbon and Materials Processing Group (2009), Cambridge University, Department of Engineering, Cambridge, U.K., http://www.lcmp.eng.cam.ac.uk/welcome/equipment/flexible-asymmetric-spinning.
119. Schultz, H. (1993) *Electron Beam Welding*, Woodhead Publishing/The Welding Institute, Cambridge, England.
120. Nakamura, A., Asano, H., Kawakami, S., Takashi, I., Takashi, F. (2010) Full-scale test of over pack closure techniques for high level waste repository operation—Welding methods and UT Systems for long-term structural integrity of weld joints, *Proceedings of the International Conference on Radioactive Waste Management and Environmental Remedi*, Vol. 2, pp. 375–382.
121. Dey, V., Pratihar, D.K., Datta, G.L., Jha, M.N., Saha, T.K., Bapat, A.V. (2009) Optimization of bead geometry in electron beam welding using a genetic algorithm, *Journal of Materials Processing Technology*, 209(3): 1151–1157.
122. Sun, Z., Karppi, R. (1996) The application of electron beam welding for the joining of dissimilar metals: An overview, *Journal of Materials Processing Technology*, 59(3): 257–267.
123. SubsTech Substances & Technologies (2007), http://www.substech.com/dokuwiki/doku.php?id=electron_beam_welding_ebw.
124. Asano, H., Maeda, K., Aritomi, M. (2006) Long-term integrity of waste package final closure for high level waste geological disposal, applicability of TIG welding method to overpack final closure, *Journal of Nuclear Science and Technology*, 43: 206–221.
125. Watkins, A.D., Mizia, R.E. (2011) Optimizing long-term stainless steel closure weld integrity in DOE standard spent nuclear canisters, ASM International, pp. 424–426.
126. Weman, K. (2003) *Welding Processes Handbook*, CRC Press LLC, New York.
127. Wu, C.S., Ushio, M., Tanaka, M. (1997) Analysis of the TIG welding arc behavior, *Computational Materials Science*, 7(3): 308–314.
128. Euroweld: Used Welding and Manipulation Equipment (n.d.), Euroweld: Your partner in professional used welding equipment, GS Oud Gastel, the Netherlands, http://usedwelding-equipment.com/tig-welding/.
129. Panigrahi, S.K., Sarangi, N., Chandrasekhar, U. (2015) Experimental evaluation of overload capability of an annular combustor casing of a gas turbine engine, *Journal of Experimental Techniques*, 40(2): 841–848.

130. Military Standard, MIL-E-005007E(AS), September 1, 1983, p. 84.
131. Wilkinson, C., Gostelow, C.R. (1994) High temperature cyclic behavior of aerospace materials: Room temperature validation tests of Ti–6Al–4V, Advisory Group for Aerospace Research and Development (AGARD) advisory report 328, North Atlantic Treaty Organisation (NATO).
132. Munoz, V., Rohr, G.A., Cavalieri, A.L., Tomba, M. (2012) Experimental procedure for the mechanical evaluation of oxide-carbon refractories by strain measurement, *Journal of Testing and Evaluation*, 40: 194–201.
133. Okayasu, M., Sato, Y. (2012) New experimental techniques for in-situ measurement of the damage characteristics of piezoelectric ceramics under high cycle fatigue testing, *Journal of Experimental Mechanics*, 52: 1009–1020.
134. Sundaram, A.B., Kesavan, K., Parivallal, S., Ravisankar, K. (2011) Technique for instrumentation and measurement of interfacial strains in FRP strengthened concrete structures using FBG sensors, *Journal of Experimental Techniques*, 8: 35–38.
135. ASTM Standard E-8 (2008) Standard test methods for tension testing of metallic materials, ASTM Internationals.
136. GTRE Internal Report No. GTRE/SIMA/MG/MTL/10/02-001, February 2010.
137. Holman, J.P. (2002) *Heat Transfer*, 9th edn., Mc-Graw Hill Book Company, pp. 1–26.
138. Wang, C.C., Choiu, C.B., Lu, D.C. (1996) Single-phase heat transfer and flow friction correlations for micro fin tubes, *International Journal of Heat and Fluid Flow*, 17: 500–508.
139. Winterton, R.H.S. (1998) Technical notes—Where did the Dittus and Boelter equation come from, *International Journal of Heat Mass Transfer*, 41(4–5): 809–810.
140. Rathod, A. (2004) Kaveri engine combustor casing thermal validation, GTRE Internal Report No. HTG/0801/43, Gas Turbine Research Establishment.
141. Fuente, E. (2009) Von mises stresses in random vibration of linear structures, *Computers & Structures*, 87(21–22): 1253–1262.
142. Eraslan, A.N. (2002) Von mises yield criterion and nonlinearly hardening variable thickness rotating annular disks with rigid inclusion, *Mechanics Research Communications*, 29(5): 339–350.
143. Brock, L.M. (1996) Effect of thermoelasticity and a Von mises condition in rapid steady-state quasi-brittle fracture, *International Journal of Solids and Structures*, 33(28): 4131–4142.
144. Boumedine, H., Coujou, A., Grilhe, J. (1996) Von mises stress for two-dimensional elasticity problems, *Scripta Materialia*, 34(3): 401–406.
145. Sova, J.A., Crews, J.H. Jr., (2004) A method for determining local elastplastic stress and strain in metallurgically bonded nothed laminates subjected to a loading cycle, NASA Technical Note, NASA TN D-7766, National Aeronautics Space Administration.
146. Jouve, D. (2013) Analytic study of plastic necking instabilities during plane tension tests, *European Journal of Mechanics A/Solids*, 39: 180–196.
147. Hoang, T.D., Herbelot, C., Imad, A., Benseddiq, N. (2013) Numerical modelling for prediction of ductile fracture of bolted structure under tension shear loading, *Finite Elements in Analysis and Design*, 67: 56–65.
148. Moreno, V., Meyers, G.J., Kaufman, A., Halford, G.R. (1983) Nonlinear structural and life analysis of a combustor liner, *Computers & Structures*, 16(1–4): 509–515.

149. Adasooriya, N.D., Siriwardane, S.C., Ohga, M. (2013) A simplified approach to predict the failure of steel members under interaction effect of fracture and fatigue, *International Journal of Fatigue*, 47: 161–173.
150. Graça, M., Duarte, A., Coelho, P.J., Costa, M. (2013) Numerical simulation of a reversed flow small-scale combustor, *Fuel Processing Technology*, 107: 126–137.
151. Chowdhury, S.G., Mukhopadhyay, N.K., Das, G., Das, S.K., Bhattacharya, D.K. (1998) Failure analysis of a weld repaired steam turbine casing, *Engineering Failure Analysis*, 5(3): 205–218.
152. Lin, S.H., Pan, J., Tyan, T., Prasad, P. (2003) A general failure criterion for spot welds under combined loading conditions, *International Journal of Solids and Structures*, 40(21): 5539–5564.
153. Gary Marquis, G. (2010) Failure modes and fatigue strength of improved HSS welds, *Engineering Fracture Mechanics*, 77(11): 2051–2062.
154. Schneider, F., Jones, N. (2003) Influence of spot weld failure on crushing of thin walled structural sections, *International Journal of Mechanical Sciences*, 45(12): 2061–2081.
155. Dziubinski, J. (1991) Fatigue failure criterion based on plastic strain energy density applied to welds, *International Journal of Fatigue*, 13(3): 223–226.
156. Zappalorto, M., Paolo Lazzarin, P. (2009) A new version of the Neuber rule accounting for the influence of the notch opening angle for out-of-plane shear loads, *International Journal of Solids and Structures*, 46(9): 1901–1910.
157. Ye, D., Matsuoka, S., Suzuki, N., Maeda, Y. (2004) Further investigation of Neuber's rule and the equivalent strain energy density (ESED) method, *International Journal of Fatigue*, 26(5): 447–455.
158. Brechet, Y., Magnin, T., Sornette, D. (1992) The Coffin–Manson law as a consequence of the statistical nature of the LCF surface damage, *Acta Metallurgica et Materialia*, 40(9): 2281–2287.
159. Huang, J.H., Si, Y., Zheng, L.G., Dong, X.H. (1992) A dislocation model of low-cycle fatigue damage and derivation of the Coffin–Manson equation, *Materials Letters*, 15(3): 212–216.
160. Zheng, C.X., Lei, S.H. (2006) Research on bursting pressure formula of mild steel pressure vessel, *Journal of Zhejiang University Science A*, 7(Suppl. II): 277–281.
161. Rajan, K.M., Deshmande, P.U., Narasimhan, K. (2011) Experimental studies on bursting pressure of thin walled flow formed pressure vessels, *Journal of Materials Processing Technology*, 125–126: 228–234.
162. George, F.V.V. (2004) *Metallography and Microstructures, ASM Metals Handbook*, Vol. 9, ASM International, Materials Park, OH, p. 888.
163. Fengjun, L., Quan, L., Guoru, F. (2010) Failure analysis of an aero-engine combustor liner, *Engineering Failure Analysis*, 17(5): 1094–1101.
164. Detlef, G.D., Friedrich, W.A., Troger, H.D., Kleemann, W.J. (1999) Unexpected findings in the investigation of an airplane crash, *Forensic Science International*, 104(2–3): 189–194.
165. Sarangi, N., Panigrahi, S.K., Patel, S.K., Ramesh, P., Ganeshachar, M.D. (2013) Effect of fatigue on proof strength of gas turbine combustor casing, *Journal of Failure Analysis and Prevention*, 13(1): 102–111.

Index

A

Adaptors
- assembly arrangement, 61–62
- bottom fixing plate, 62–63
- bottom sealing drum, 62–63
- cross-section of, 61
- inner bottom fixing plate, 63–64
- seals, 68
- special studs, 67
- special tie rod, 64
- top fixing plate, 65–66
- top holding plate, 66–67
- top sealing plate, 65
- top support plate, 65–66

Aero gas turbine engine
- axial flow and centrifugal compressor combination, 8–9
- axial flow compressor, 7–8
- axial flow turbine, 11–12
- centrifugal compressor, 7–8
- combustion chamber
  - annular combustion chambers, 16–18
  - basic requirements, 10
  - description, 10
  - design complexities, 17–19
  - materials, 19–20
  - multiple combustion chambers, 15–16
  - tubo-annular combustion chambers, 16–17
- design philosophy, 14–15
- development, 4–6
- multi-spool compressor, 9
- propelling nozzle, 12–13
- single spool compressor, 9
- working principle, 6–7

Airworthiness, 55

Annular combustors, 23–24

Axial flow compressor, 7–8

Axial flow turbine, 11–12

C

Centrifugal compressor, 7–8

Certification and acceptance, 139

Coffin–Manson equation, 94–95

Cold forging, 102–103

Combustion chamber
- annular combustion chambers, 16–18
- basic requirements, 10
- description, 10
- design, 23–25
- design complexities, 17–19
- materials, 19–20
- multiple combustion chambers, 15–16
- tubo-annular combustion chambers, 16–17

Combustor rig tests, 24

Combustor's design validation test, 1

Component testing, 54

Configuration of casing, 75–76

Crack modeling method, 24

Cycle counting methods
- fatigue crack growth analysis, 33
- fatigue-meter count method, 31
- Inconel 718 specimen, 33–34
- level-crossing count method, 30–31
- mean-crossing peak count method, 30
- peak count method, 30
- range count method, 31
- range-mean count method, 31
- range pair method, 31–32
- rules, 32–33

D

Dittus Boelter equation, 87–90

Dye penetrant inspection, 44–45

E

Eddy current inspection technique, 47–49
Electron beam welding method, 72–74
Engine cyclic endurance tests, 24

F

Failure mode and effect analysis, 1
Fatigue design philosophy
crack initiation period, 28
crack nucleation, 27
crack propagation period, 28
cycle counting methods
fatigue crack growth analysis, 33
fatigue-meter count method, 31
Inconel 718 specimen, 33–34
level-crossing count method, 30–31
mean-crossing peak count method, 30
peak count method, 30
range count method, 31
range-mean count method, 31
range pair method, 31–32
rules, 32–33
damage assessment, 37–39
damage causes, 26
definition, 25
graphical representation, 27
HCF regime, 28
life evaluation methods, 34–37
local strain method, 25
macrocracks, 26
microcracks, 26
microscopic plastic strains, 26
S-N curve, 25
Southwest Flight 812, 28–29
spectrum loading, 29
strength, welded structures, 24
thermo-mechanical fatigue characteristics, 25
Fatigue life evaluation
inspection methodology, 79–81
instrumentation, 77–79
universal slope equation, 95
Fatigue loaded diffuser casing
metallographic techniques, 119–126
post-fracture investigation
microhardness measurement, 131–132
optical microscopy, 131–132, 137
SEM, 129, 131–135, 137
SZM, 128–131, 137
visual observation, 128
proof pressure capability
experimental procedure, 127
test component details, 126
Fatigue-meter count method, 31
Finite element method, 82–84
Fluorescent particle inspection (FPI), 110, 118
Forging process
cold forging, 102–103
description, 101
forged casing, 105
hot forging, 102–103
steam hammers, 102
warm forging, 102

H

High-cycle fatigue (HCF) regime, 28
Hot forging, 102–103
Hydrostatic testing, 113

I

Igniter boss
boss-shell zone, 91, 92
Coffin–Manson equation, 94–95
configuration of, 99
geometric details, 89, 91
LCF life, 96–97
linear elastic stress analysis, 93
Neuber curve, 94, 98
solid model, 89, 91
von Mises stress distribution, 93–95
Inconel 718, 81–83

L

Leak before burst, 114
Level-crossing count method, 30–31
Liquid penetrant inspection, *see* Dye penetrant inspection
Lowcycle fatigue (LCF), 24

M

Magnetic particle inspection (MPI), 42–44
Manufacturing methods
  electron beam welding method, 72–74
  metal spinning process, 70–72
  TIG welding, 74–75
Mean-crossing peak count method, 30
Mechanical testing, 53
Metallographic techniques
  description, 120
  optical microscopy, 125–126
  sample preparation, 120
  SEM, 121–125
  SZM, 120–122
Metal spinning process
  advantages, 71–72
  disadvantages, 72
  operations sequence, 70–71
  schematic diagram, 70–71
  spin chuck, 70
  steel tooling, 70
  wood tooling, 70
Microcracks, 26
Military standards, 141
Miner's rule, 34–36
Modified combustor casing
  assembly and trial run, 106–108
  cyclic pressure test, 109
  fatigue life test, 109
  forging process
    cold forging, 102–103
    description, 101
    forged casing, 105
    hot forging, 102–103
    steam hammers, 102
    warm forging, 102
  FPI reports, 110
  instrumented combustor casing, 106, 108
  manufacturing plan, 103–104
  PWHT, 105
  safety test, 113–118
  strain gauges, 106–107
  vs. version 1 combustor casing, 110
Multi-spool compressor, 9

N

NDT, *see* Nondestructive testing methods
Neuber's methodology, 94
Neuber's rule, 36–37
Nondestructive testing methods (NDT)
  dye penetrant inspection, 44–45
  eddy current and electromagnetic inspection, 47–49
  magnetic particle inspection, 42–44
  radiography inspection, 40–42
  ultrasonic inspection, 45–47
Numerical analysis
  boundary conditions, 85
  Dittus Boelter equation, 87–90
  finite element method, 82–84
  igniter boss
    boss-shell zone, 91, 92
    Coffin–Manson equation, 94–95
    configuration of, 99
    geometric details, 89, 91
    LCF life, 96–97
    linear elastic stress analysis, 93
    Neuber curve, 94, 98
    solid model, 89, 91
    von Mises stress distribution, 93–95
  Nusselt equation, 87
  pressure load, 85
  thermal load, 85–86
Nusselt equation, 87

O

Optical microscopy
  application, 125–126
  description, 125
  post-fracture investigation, 132, 135–137

P

Peak count method, 30
Postweld heat treatment (PWHT), 105
Pressure vessels design, 113
Pressure vessel standards, 114

Proof pressure test
  burst pressure calculation, 113
  description, 114
  FPI reports, 118
  pressure cycle, 117
  set up, 115–116
  strain cycles, 117
  test component details, 114–115

R

Radiography inspection, 40–42
Range count method, 31
Range-mean count method, 31
Range pair method, 31–32

S

Scanning electron microscopy (SEM)
  applications, 123
  components, 123–125
  description, 121–122
  post-fracture investigation
    crack origination, 137
    energy dispersive x-ray analysis, 131, 135
    fractography, 129, 131–134
  qualitative and quantitative analysis, 122
Seam welding technology, 36
SEM, *see* Scanning electron microscopy
Single spool compressor, 9
Standard material characterization tests, 24
Stereo zoom microscopy (SZM), 121
  applications, 121
  description, 121
  post-fracture investigation, 128–131, 137
Structural integrity, 53
Structural test, 2
SZM, *see* Stereo zoom microscopy

T

Test facilities, 1
  hydraulic power supply, 57–58
  pressure test facility, 56–57
  pressure transducer, 59–60
  safety relief valve, 59
  schematic diagram, 56–57
  servo controller, 59
  servo valves, 58
  subsystems, 60–61
Trent engine, 23
Tungsten inert gas (TIG) welding, 74–75, 137

U

Ultrasonic inspection, 45–47

V

von Mises stress distribution, 93–95

W

Warm forging, 102
Weld factor, 36

For Product Safety Concerns and Information please contact our EU representative GPSR@taylorandfrancis.com Taylor & Francis Verlag GmbH, Kaufingerstraße 24, 80331 München, Germany

Batch number: 10726532

Printed by Printforce, the Netherlands